T0181189

Studies in Big Data

Volume 2

Series Editor

Janusz Kacprzyk, Warsaw, Poland

For further volumes:
http://www.springer.com/series/11970

Mohamed Medhat Gaber · Frederic Stahl
João Bártolo Gomes

Pocket Data Mining

Big Data on Small Devices

 Springer

Mohamed Medhat Gaber
School of Computing Science
 and Digital Media
Robert Gordon University
Aberdeen
UK

João Bártolo Gomes
Institute for Infocomm Research (I^{2}R)
Agency for Science, Technology
 and Research (A*STAR)
Singapore

Frederic Stahl
School of Systems Engineering
The University of Reading
Reading
United Kingdom

ISSN 2197-6503   ISSN 2197-6511   (electronic)
ISBN 978-3-319-34686-1   ISBN 978-3-319-02711-1   (eBook)
DOI 10.1007/978-3-319-02711-1
Springer Cham Heidelberg New York Dordrecht London

Printed on acid-free paper

Springer is part of Springer Science+Business Media (www.springer.com)

Acknowledgements

The authors would like to acknowledge the work done by our visiting students at the University of Portsmouth *Victor Mandujano* and *Oscar Campos* in porting the *PDM* system from the desktop to the mobile environment. Also their contribution to Chapter 4 is acknowledged.

The authors also acknowledge *Professor Philip S Yu*, *Professor Max Bramer* and *Professor Ernestina Menasalvas* for contributing to discussions regarding the development of this project.

Finally, we would like to thank our students at the University of Portsmouth that helped with performing the experimental study reported in Chapter 3, namely, *Paul Aldridge*, *David May*, and *Han Liu*.

Contents

Chapter 1
Introduction

Thanks to continuing advances in mobile computing technology, it has become possible to perform complex data processing on-board small handheld devices. It has taken over a decade from moving to handheld devices that can place and make calls, send short message, organize the calendar and save some data, to handheld devices that are as powerful as our ten year old computer servers. It has never been a better time to exploit the increasing computational power of small handheld devices like smartphones and tablet computers. The market studies have shown that tablet computers will soon replace our laptop computers. In a recent market study in the USA, it has been found that 46% of Americans think that tablets will replace laptops[1].

Among the complex processing tasks that our small devices became capable of *knowledge discovery* and *data mining*. Historically, the two terms *knowledge discovery* and *data mining* were different [51]. Data mining used to be the main stage of the iterative process of *knowledge discovery* , as shown in Figure 1.1. However, the two terms are used interchangeably nowadays. *Data mining* could be defined as the nontrivial process of extracting interesting knowledge structures from typically large data repositories. From this definition, typical data processing tasks from storage to retrieval are not considered to be *data mining*. Furthermore, the knowledge structures to be mined need to be of interest to the user. Knowledge structures could be typically subdivided to patterns and data models. Finally, *data mining* techniques are typically applied on large data repositories like very large databases and data warehouses.

1.1 Introduction to Mobile Data Mining

More and more data mining applications are running on mobile devices such as 'Tablet PCs', smartphones and Personal Digital Assistants (PDAs). The ability

---

[1] http://mashable.com/2012/01/10/tablets-laptops-study/

M.M. Gaber, F. Stahl, and J.B. Gomes, *Pocket Data Mining*, Studies in Big Data 2,
DOI: 10.1007/978-3-319-02711-1_1, © Springer International Publishing Switzerland 2014

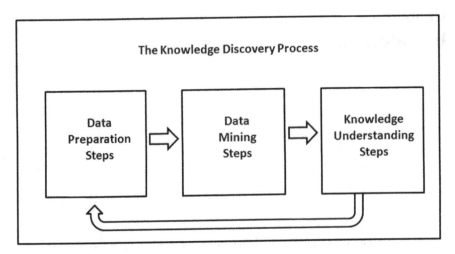

Fig. 1.1 The Knowledge Discovery Process

to make phone calls and send SMS messages nowadays seems to be merely an additional feature rather than the core functionality of a smartphone. Smartphones offer a wide variety of sensors such as cameras and gyroscope as well as network technologies such as *Bluetooth*, and *Wi-Fi* with which a variety of different data can be generated, received and recorded. Furthermore smartphones are computationally able to perform data analysis tasks on these received, or sensed data such as data mining. Many data mining technologies for smartphones are tailored for data streams due to the fact that sensed data is usually received and generated in real-time, and due to the fact that limited storage capacity on mobile devices requires that the data is analyzed and mined on the fly while it is being generated or received. For example, the *Open Mobile Miner (OMM)* tool [62, 49] allows the implementation of data mining algorithms for data streams that can be run on smartphones. *OMM* has been designed and built around a suite of data stream mining techniques that Gaber et al [35, 49] have proposed and implemented. A unique feature of *OMM* techniques is the adaptability to availability of computational resources of the computational device. This adaptation has been based on the theoretical framework provided by the *Algorithm Granularity* proposed by Gaber [34]. Figure 1.2 shows the graphical user interface of *OMM*.

Earlier work to *OMM* targeting mining of data streams on-board smartphones has been reported, including *MobiMine* [58] and *VEDAS* [56] and its commercial version *MineFleet* [59, 9]. These systems will be further investigated in Chapter 2 of this monograph.

However, to the best of our knowledge, all existing data mining systems for mobile devices either facilitate data mining on a single node or follow a centralized approach where data mining results are communicated back to a server which makes decisions based on the submitted results. Constraints that require the distribution of data mining tasks among several smartphones are, large and fast data

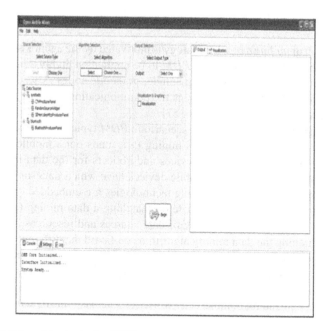

Fig. 1.2 OMM used in Analyzing ECG

streams, subscription fees to data streams, and data transmission costs in terms of battery and bandwidth consumption. The data transmission cost can be lowered by processing parts of the same data stream locally on different smartphone devices that collaborate only by exchanging local statistics, or locally generated data mining models rather than raw data. The collaborative data mining on smartphones and 'Tablet PCs', facilitated by building an ad hoc network of mobile phones, will allow to build significantly useful analysis tasks, however, this area remains widely unexplored. *Pocket Data Mining (PDM)* aims to explore the full potential of such an open area. In the following, an overview of *PDM* will be provided to the reader. A full treatment of the *PDM* framework is provided in sufficient details in Chapter 3.

1.2   Pocket Data Mining: An Overview

In this monograph, we describe and evaluate the *Pocket Data Mining (PDM)* framework, coined and proven to be computationally feasible in [92]. *PDM* has been built as a first attempt to explore collaborative and distributed data mining using stream mining technologies [41], mobile software agents technologies [106, 71] and programming for mobile devices such as smartphones and 'Tablet PCs'. The main motivation in developing *PDM* is to facilitate the seamless collaboration between users of mobile phones which may have different data, sensors and data mining technology available.

The adoption of *mobile agent* technology is motivated by the agent's autonomous decentralized behavior, which enables *PDM* to be applied on highly dynamic problems and environments with a changing number of mobile nodes. A second motivation for using *mobile agent* technology is the communication efficiency of mobile agent based distributed data mining [84, 57].

A general agent based collaborative scenario in *PDM* typically has the following steps. A mobile device that has a data mining task sends out a mobile agent that roams the network of other mobile devices and collects for the data mining task useful information, such as which mobile devices have which data sources and/or sensors available and which data mining technologies are embedded on these devices. Following this, a decision is made on matching a data mining task to each mobile device according to availability of data sources and resources. The task is then initiated running the data mining algorithms on-board the different mobile devices. Whenever needed, a mobile device can send a different type of mobile agent to consult the other mobile devices on a data mining result. This process is further detailed in Chapter 3.

The growing demand for commercial *knowledge discovery* and *data mining* techniques has led to an increasing demand of classification techniques that generate rules in order to predict the classification of previously unseen data. Hence classification rule induction is also a strong candidate technology to be integrated into the *PDM* framework. For example, the classification of data streams about the stock market could help brokers to make decisions whether they should buy or sell a certain share. Also in the area of health care, classification of streaming information may be beneficial. For example, the smartphone may record various health indicators such as the blood pressure and/or the level of physical activity and derive rules that may indicate if a patient needs medical attention, the urgency and what kind of medical attention is needed. For this reason, a version of *PDM* that incorporates two widely acceptable data stream classification techniques, namely, *Naive Bayes* and *Hoeffding Trees*, has been created and is evaluated in this book.

1.3   Monograph Structure

This monograph style book is organized as follows. Chapter 2 provides the required background for the areas that contribute to the *PDM* framework. These areas include mobile data mining, mining data streams, high performance data mining and mobile agent technologies. After setting the scene up in Chapter 2, Chapter 3 provides the details of our *PDM* framework, including implementation and evaluation of the performance. Porting the implementation to the *Android* mobile platform witnessed experiences in software development, that is worth reporting in Chapter 4. Chapter 5 extends the basic implementation of *PDM* by incorporating contextual information, while the experimental evaluation of this context-aware extension is detailed in

Chapter 6. Chapter 7 provides insights of the different applications that can benefit from the *PDM* framework, providing data analytics practitioners with fresh ideas on how *PDM* can be applied in a variety of fields. Finally, the monograph is concluded with a summary and pointers to future directions in research in Chapter 8.

Readers can find it easy to read the monograph, as the chapters are loosely coupled. This gives flexibility and ease of readership. However, we advise the reader to go through all the chapters for the benefit of acquiring the theoretical and practical background needed to fully comprehend and appreciate what *PDM* can offer.

Chapter 2
Background

A number of developments in areas related to *Big Data* analytics[1] have led to the realization of our *PDM* framework reported in this monograph. Mobile data mining, data stream mining, and parallel and distributed data mining and mobile software agents are four areas that collectively contributed to the possibility of developing *PDM*. Mobile software agents as a flexible software development framework allowed *PDM* to have on the fly deployment of data mining techniques, if needed. Based on these areas, the four pillars for *Pocket Data Mining* can be depicted as in Figure 2.1. These areas will be reviewed in this chapter.

2.1 Data Mining on Mobile Devices

Data mining on mobile devices can be broadly classified into two categories: (1) *mobile interface*; and (2) *on-board execution*. In the former one, the mobile device is used as an interface to a data mining process that runs on a high performance computational facility, like the *cloud*. This represented the early history in the area of mobile data mining, with the *MobiMine* system [58] to analyze stock market share prices being the first realization of this category. In the on-board execution category, the mobile device is used to not only set the parameters and visualize the results, but also to run the data mining process. This category has been the result of the continuous advances in mobile devices like smartphones and tablet computers. Many of these devices are comparable in terms of computational power to computer servers that run data-intensive tasks a decade ago. In the following, we give representative examples of each of the two categories.

2.1.1 Mobile Interface

As aforementioned, *MobiMine* [58] has marked the first realisation of a mobile data mining system. The system paid special attention to how efficiently a decision tree

---

[1] http://www-01.ibm.com/software/data/bigdata/

M.M. Gaber, F. Stahl, and J.B. Gomes, *Pocket Data Mining*, Studies in Big Data 2,   7
DOI: 10.1007/978-3-319-02711-1_2, © Springer International Publishing Switzerland 2014

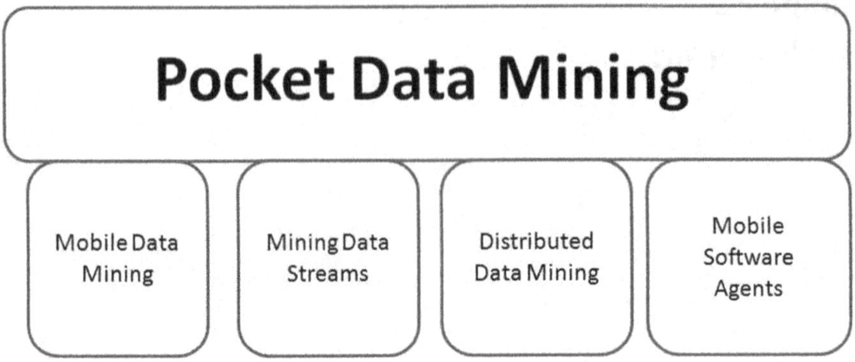

Pocket Data Mining

Mobile Data Mining	Mining Data Streams	Distributed Data Mining	Mobile Software Agents

Fig. 2.1 PDM Pillars

can be communicated over a channel with limited bandwidth; a real problem at the time, that now seems not to be of an issue. The motivation behind the development of *MobiMine* is to help businessmen on the move to analyze the stock market shares and take a decision without having to attend to their personal computers. A breakdown of tasks between the Personal Digital Assistant (PDA) and the server was designed - such that all the computationally intensive processes are performed at the server side. To ensure that communication will not fail the system, a Fourier transformation of the signal was used. *MobiMine* will be given special treatment in this book as a potential application area for *PDM*. Readers are referred to Chapter 7 for this discussion.

At the time of development, *MobiMine* was a pioneering piece of work. Despite the fact that this client/server architecture with a thin client representing the mobile device was motivated by the limited power of such devices at the time, it has opened the door for further development that exploited the continuous advances in hardware of the handheld devices.

2.1.2   On-board Execution

Not long after the short era of the *MobiMine* system, the same group at the *University of Maryland, Baltimore County (UMBC)* led by *Kargupta* developed the *Vehicle Data Stream Mining System (VEDAS)* [56] . It is a ubiquitous data stream mining system that allows continuous monitoring and pattern extraction from data streams generated on-board a moving vehicle. The mining component is located on a PDA placed on-board the vehicle. *VEDAS* uses online incremental clustering for modelling of driving behaviour, thus detecting abnormal driving behaviour.

VEDAS has opened the door to a new area of development, considering handheld devices as portable computational power, rather than them being regarded as limited

purpose devices of making and receiving calls, and sending and receiving short
messages. At the time of *VEDAS*, limited computational resources and connectivity
of handheld devices have been the major issues to be addressed to realise the full
potential of what has been termed as *Mobile Data Mining (MDM)* .

MDM has been tightly coupled with the generic area of mining data streams,
discussed briefly in the following section. This association is due to the fact that
data received or synthesized on-board mobile devices typically arrive as a stream.

2.2   Data Mining of Streaming Data

Mining data streams, on the other hand, is a more recent topic of research in the area
of Big Data analytics. A concise review of the area by Gaber et al is given in [41].
A more detailed review is given in [42]. The area is concerned with analysis of data
generated in a high speed relative to the state-of-the-art computational power, with
a constraint of real-time demand of the results. Hundreds of algorithms have been
proposed in the literature addressing the research challenges of data stream mining. Notable success of the use of *Hoeffding bound* to approximate the data mining
models for streaming data has been recognized [32]. The two-stage process of online summarisation and offline mining of streaming data, proposed by Aggarwal el
al [7, 8], has been also recognized as a feasible approach to tackle the high data rate
problem.

The resource constraints and high speed aspects of data stream mining have been
addressed by Gaber et al [34, 39, 40] by proposing the *Algorithm Granularity* approach (See Figure 2.2). The approach is generic and could be plugged into any
stream mining algorithm to provide resource-awareness and adaptivity. In Figure
2.2, the *AIG* stands for *Algorithm Input Granularity* and is used to adapt the mining
algorithm from the input side preserving the battery charge prolonging its lifetime.
Typical techniques for *AIG* include sampling and load shedding. Also in the same
figure, *APG* stands for *Algorithm Processing Granularity* and is concerned with result approximation that makes the mining algorithm require less or more processing
cycles for the same data size. A typical example in this category is randomization
techniques. Finally, the *AOG* stands for *Algorithm Output Granularity* and works
on controlling the size of the algorithm output by changing the algorithm settings
accordingly.

Mining data stream algorithms is at the heart of our *Pocket Data Mining* framework. Typically, *PDM* utilizes streaming algorithms on-board smartphones captur-
ing patterns generated by the many sources of data stream our smartphones receive
and generate.

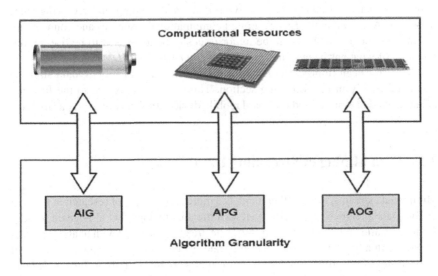

Fig. 2.2 Algorithm Granularity Approach

2.3 Parallel and Distributed Data Mining

Big Data in the literal sense of very large datasets challenge our computational hardware. Businesses as well as science are challenged by these massive and potentially distributed data sources. For example, in the area of molecular dynamics, a sub field of bioinformatics, data is generated on a large scale. Experimental molecular dynamics simulations produce massive amounts of data that challenge data management, storage and processing hardware to analyse the simulations' data [17, 86]. The NASA's system of *Earth Orbiting Satellites (EOS)* and other space borne probes [101] is a further example of massive data generation. *EOS* is still ongoing and produces approximately one terabyte of data per day. There are also many examples in the business world according to [4]. In [4], it is claimed that the two largest databases of *Amazon* comprise more than 42 terabyte of data combined as of 2012. Two well known astronomy data bases, the *GSC-II* [67] and the *Sloan survey* [97] hold data in the range of terabytes. However, it is not only the number of data records that are held in a database that contribute to the explosion of data, but also the size of each individual record in terms of number of data features (also known as attributes), that challenge our computer hardware. In the context of 'market basket' analysis this may very well comprise as many features as there are items in a supermarket, which could potentially be tens of thousands. In general the digital universe is growing at a fast pace, for example the authors of [43] predict the size of the digital universe in the year 2020 to be 44 times as big as it was in the year 2009.

It is not obvious how parallel data mining techniques fit in with *Pocket Data Mining*, as it is unlikely that smartphones may be used to process massive datasets. It would seem more likely that for data mining of massive datasets, super computer architectures of parallel machines or loosely coupled networks of PCs are used rather than smartphones. However, if one considers that the processing capabilities of smartphones are still limited, it seems obvious that even modest sized datasets may impose a computational challenge. Hence parallel data mining techniques should be considered within the *Pocket Data Mining* framework. Also as the data sensing capabilities improve, it can be expected that the data generated on smartphone devices will increase as well in the future. Hence Section 2.3.1 outlines a few basic principles of parallel data mining to keep in mind for future *PDM* architectures. A more comprehensive survey of parallel data mining techniques can be found in [88].

Computational constraints for data mining of datasets on smartphones is not the only challenge. As smartphones have sensing capabilities, networks where distributed datasets and data streams exist are likely. Thus data fusion techniques may be used. However, the problem that arises here is that the data may be too large to be transmitted among mobile devices, this may be due to bandwidth or battery power constraints. However, also geographically distributed data needs to be analysed and data mined. Similar problems exist in traditional data mining (data mining without mobile devices), for example in [17] researchers address the issue of distributed datasets in the area of molecular dynamics simulations. Different molecular dynamics research groups want to share data they generated locally in different simulations, but the size of the data is too large to be transmitted. However, techniques exist to store and analyze such geographically distributed datasets and *Pocket Data Mining* should utilize these techniques, as transmission of data from smartphones is expensive in terms of battery consumption. Section 2.3.2 outlines a few basic principles of geographically distributed data mining techniques to keep in mind for future *PDM* architectures. A more comprehensive survey of distributed data mining techniques can be found in [74, 91]. Section 2.3.3 highlights how Parallel and Distributed Data Mining techniques are in line with PDM.

2.3.1   Parallel Data Mining

Multiprocessor computer architectures are commonly categorized into 'tightly-coupled' and 'loosely-coupled' architectures.

- In *tightly-coupled* multiprocessor architectures all processors use the same shared memory, hence distributing data is not required. Such *tightly-coupled* multi-processor architectures are also often referred to as '*Shared memory Multiprocessor machines*' (SMP).
- In a *loosely-coupled* multiprocessor architecture each processor uses its own private memory. They are often a collection of several individual computers. In data driven applications, using a *loosely-coupled* architecture, a distribution of the data is often required together with communication of data and accumulation mechanisms. Such *loosely-coupled* architectures are also often referred to as '*Massively Parallel Processors*' (MPP).

Both types of multiprocessor architectures are illustrated in Figure 2.3 [85]. The tighly-coupled architecture is using a memory bus in order to access and share data in the (shared) memory; and the loosely-coupled architecture is using a communication network in order to share data from individual memories. *PDM* is naturally a loosely coupled system, as it consists of mobile devices that are interconnected using a communication network, hence tightly coupled systems are not discussed further in this book.

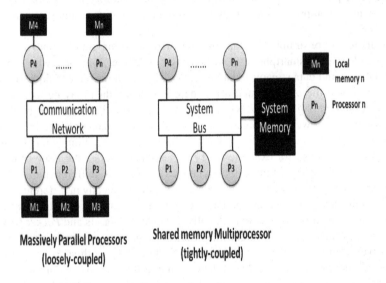

Massively Parallel Processors (loosely-coupled)

Shared memory Multiprocessor (tightly-coupled)

Fig. 2.3 The left hand side of the figure shows a tightly-coupled multiprocessor computer architecture and the right hand side a loosely-coupled multiprocessor computer architecture

There are two general ways in which multiprocessor architectures are used in data mining in order scale up data mining tasks to large volumes of data, *task* and *data parallelism*. A data mining algorithm is task parallel if its program code is executed concurrently on several processors. Task parallelism is also referred to *control parallelism* in the data mining literature [33]. Compared with *task parallelism*, in *data parallelism* the same task (program code) is executed concurrently on several subsets of the dataset on several processors [52]. *Data parallelism* is often the method of choice for scaling up data mining tasks as the computational workload of data mining tasks is directly dependent on the amount of data that needs to be processed.

Data parallel data mining algorithms are sometimes referred to as 'distributed data mining' algorithms, because of the fact that the data is distributed to several computing nodes [88]. This is confusing as 'distributed data mining' also refers to the mining of geographically distributed data sources which is in contrast to data parallel data mining algorithms not necessarily concerned with achieving a shorter execution time due to data parallelization [88]. In this book, we use distributed data mining in reference to mining of geographically distributed datasets and data streams as described in Section 2.3.2 and not data parallel data mining algorithms.

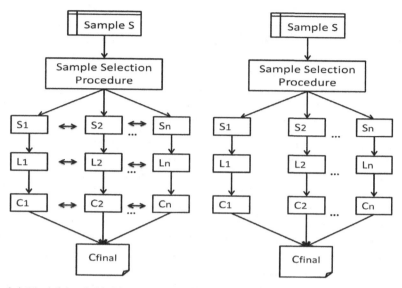

Fig. 2.4 The left hand side illustrates the Cooperating Data Mining Model and the right hand side the Independent Multi Sample Mining Model

In [78], several data parallel data mining approaches were categorized into several data mining models. Two important models will be discussed here. Provost refers to the models as distributed data mining, however, in order to conform with the terminology used in this book and the fact that the models are investigated in order to scale up data mining through data distribution and concurrent execution of data mining tasks, we will call them data parallel models. All of the data parallel models described in [78] can be divided into three basic steps: the *sample selection procedure*, *learning local concepts* and combining local concepts using a *combining procedure* into a final concept description.

- *sample selection procedure*: The sample selection procedure produces n samples $S_1,..., S_n$ if there are n computing nodes available and distributes them over the n computing nodes. However, how the samples are built, is not further specified and dependent on the data mining algorithm used. Samples could be for example a subset of the instance space or a subset of the feature space. Also the size of the sub-samples is not further specified, this might be dependent on the processing power of each individual computing node or the data mining algorithm used.
- *learning local concepts*: Learning or data mining algorithms $L_1,...,L_n$ run on each of the n computing nodes. These learning algorithms learn a concept out of the

locally stored samples on each computing node. Whether these n learning algorithms $L_1,...,L_n$ do or do not communicate depends on the model. Each L derives a concept description based on the local data.

- *combining procedure*: After each L has generated its local concept description C, all Cs are communicated to one location and combined into a final concept description C_f. The combining procedure depends on the nature of the Ls. For example, each L might derive a set of candidate rules that fit the local subset of the data well; and the combining procedure then evaluates each rule how well it fits on a validation dataset. Another possibility using the cooperating data mining model is that each L generates a C that is already a part of C_f, in this case the combining procedure only assemblies the C_f using all Cs.

Provost's parallel models are very generic, they do not assume a particular data mining technique and thus fit many kinds of learning algorithms. These models should be seen as a more generic way to describe parallel and distributed data mining algorithms. Figure 2.4 depicts the *independent multi sample mining model* on the right hand side. The word 'independent' expresses that there is no communication between the computing nodes. Each L forms its concept description independently. For the independent multi sample mining model each L has only the view on the local search space reflected by the local sub-sample. However, for the cooperating data mining model, the Ls can communicate and thus gain a more global view of the data search space. This communication can be either information about local concepts, statistics about the local data or whole raw data sub-samples. However, exchanging whole sub-samples might be too expensive concerning the bandwidth consumption when dealing with very large datasets, hence communication of information about local concepts or statistics about local data is usually preferred. Some examples of algorithms that follow the *independent multi sample mining model* are the Meta Learning [25], the Random Forest [23], and the Random Prism [87] approaches, however, they do this in a sequential fashion, but could easily be executed concurrently, which has been done in some cases for example for Random Prism [95].

The probably most suitable model for parallelizing data mining algorithms is the *Cooperating Data Mining model (CDM)* depicted on the left hand side in Figure 2.4. *CDM* is based on a useful observation about certain evaluation metrics, in that every rule that is acceptable globally (according to a metric) must also be acceptable on at least one data partition on one of the n computing nodes [80, 79]. This is also known as the *invariant-partitioning property*. In *CDM* the learning algorithms cooperate in order to generate a globally acceptable concept description. A parallel classification rule induction system that follows this approach is reported in [90, 89].

2.3.2   Distributed Data Mining

In general *Distributed Data Mining (DDM)* is a term for a growing set of technologies that is concerned with the challenge of finding data mining models or patterns in geographically distributed data locations. As mentioned earlier, *DDM* is also often used as a synonym for data parallelization, however, this Section views *DDM* in the context of geographically distributed data sources.

In some applications the data is naturally distributed amongst several locations and analysts want to extract global knowledge from these distributed data sources. Such applications are used for example by large companies with geographically distributed branches that generate and manage their operational data locally. Here business analysts want to analyze all the distributed data in order to perform company wide activities such as marketing, deriving of global business strategies, assessing the company's business health, sales, etc. There are two principal ways of analyzing this data globally, the first way is to communicate the local data to a central repository and analyze it centrally, or to use *DDM* technologies in order to analyze it in a distributed way. However, the centralized analysis approach may not be feasible for several reasons, for example the data may simply be too large to transfer over the network or there may be privacy and security implications, because of which a sharing of data may not be desirable. The distributed analysis of local data sources allows to execute data mining algorithms concurrently, to combine their results and thus is speeding up the data mining process compared with the centralized approach. However, not all data mining techniques are suitable for *DDM*, some techniques suffer from data dependencies and hence involve a considerable communication overhead [24].

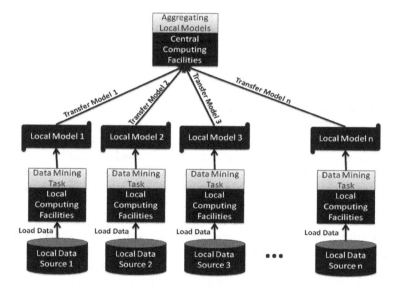

Fig. 2.5 The typical framework of a distributed data mining system. Data Mining algorithms are executed remotely on geographically distributed data sources and the generated models are aggregated on a central site.

In *DDM* techniques data processing takes usually part on a local and a global level. The typical architecture of a *DDM* technique is highlighted in Figure 2.5. The distributed data mining system distributes the data mining tasks (algorithms) to remote data locations. The algorithms are executed close to the data repositories on local computing facilities. Each algorithm generates a data mining model, for example a set of rules, and transmits these models to a central computing facility. The models are then aggregated into a global data mining model. In some applications a sub-sample of the local data sites is transmitted to the central location for validation purposes of the global model. However, these data samples need to be small in size due to bandwidth constraints.

One data mining techniques that lends itself to data mining geographically distributed datasets and to concurrent execution is ensemble learning. Figure 2.6 shows the basic ensemble learning architecture. Most ensemble learners are classifiers based on decision trees. They induce multiple base classifiers, and combine their predictions into a composite prediction, in order to improve the predictive accuracy. Such ensemble learners can easily be executed in a distributed manner by generating the base classifiers on different locations on local datasets.

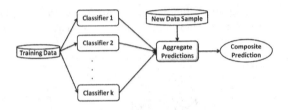

Fig. 2.6 Ensemble Learning

One of the best known ensemble learners is Breiman's previously mentioned *Random Forest (RF)* [23]. *RF* is a further development of Ho's *Random Decision Forests (RDF)* [54]. Ho proposes inducing multiple trees on randomly selected subsets of the feature space with the reasoning that traditional decision tree classifiers cannot grow over a certain level of complexity without overfitting on the training data. According to Ho the combined classification will improve as each tree will generalize better on its subset of the feature space. *RF* combines *RDF* and Breiman's *bagging* (**B**ootstrap **agg**regat**ing**) method [22]. Bagging aims to improve the classifiers predictive accuracy and to make it more stable with respect to changes in the training data. Distributed versions of *RF* exist in the context of data parallelism. For example the authors of [14] use a *Hadoop* computer cluster [2] which implements Google's MapReduce paradigm [30] for massive data parallelism in order to distribute data samples and base classifiers of *RF* evenly over a network of computers. *Hadoop* has also been used by the authors of [95] to parallelize the *Random Prism* ensemble learner.

A further ensemble learning approach is Chan and Stolfo's Meta-Learning framework [25, 26] highlighted in Figure 2.7.

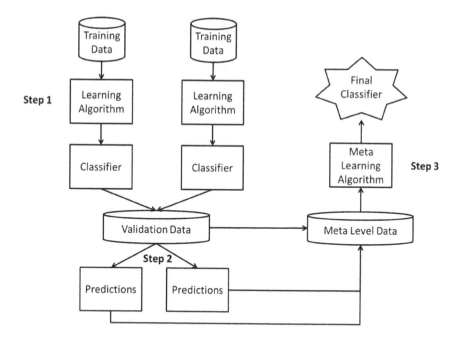

Fig. 2.7 Meta-Learning illustrated on two databases

Meta-Learning partitions the data into sub-samples and trains a classifier on each subset separately (step 1 in Figure 2.7). Next predictions are made using the trained classifiers and a validation dataset (step 2 in Figure 2.7). A Meta-Level training dataset is created from the validation data and the predictions of the base classifier on the validation data (step 3 in Figure 2.7). The base classifiers are then combined into a final classifier using various methods such as *voting*, *arbitration* and *combining* (step 3 in Figure 2.7) [25]. Recently distributed Meta-Learning technologies have been developed. For example the authors of [53] have developed a grid enabled version of the *WEKA* data mining software comprising a distributed version of the Meta-Learning approach for the execution on in the grid distributed datasets.

2.3.3   The Mobile Cloud: Pocket Data Mining in the Context of Parallel and Distributed Data Mining

It is unlikely that parallel data mining technology is used to scale up data mining to large datasets on smartphones, due to their processing, battery and memory constraints. However, because of the smartphone's limited processing capabilities, even modest sized datasets may be computationally too demanding for smartphones, hence parallelization may still be of value for data mining of moderate sized datasets on mobile devices.

Distributed computing technologies are more likely to be adopted by *PDM* in the near future, different smart devices capture different aspects of the data, hence collaborative data mining technologies such as ensemble learners that can be executed in a distributed way on a network of smartphones are most likely to be adopted by *PDM*.

If possible, even outsourcing of data mining tasks into a cloud computing environment may be considered. Mobile Cloud Computing is emerging as a new branch of Cloud Computing.

Grid computing and the recently popularized cloud computing paradigm provide an infrastructure targeted to implement computationally expensive applications such as the analysis of very large datasets. Both terms grid and cloud are often used interchangeably, and hence their usage in the literature is often blurred.

Grid computing is often explained using the analogy of an actual electrical power grid that provides electricity to the end user in a transparent way [91]. Furthermore, the end user is also able to provide electricity to other end users, for example through the usage of solar panels. According to this, grid computing provides computational resources such as processing power, storage and memory, but also allows to share the end user's own resources with other end users. Basically the grid infrastructure aims to mass resources whilst hiding their specifications underneath consistent interfaces that provide the end user with high performance and high throughput computation and storage [65].

In comparison cloud computing offers through services on demand access to applications and hardware resources over a network, such as the Internet. This sounds similar to grid computing as both paradigms aim to provide access to massive computational resources, however, cloud computing is a more commercial paradigm that offers these services through payment. In addition, cloud computing provides access to computing resources through visual tools and thus hides heterogeneity of the computational resources and their geographical distribution and faults [82].

Mobile Cloud Computing (MCC) is described by the Mobile Cloud Computing forum as an infrastructure that supports mobile applications, i.e., applications on smartphones, to move computation and data storage outside of the mobile devices into the cloud [3]. The cloud may contain powerful centralized computers that can be accessed via wireless connections on mobile devices [1]. Companies such as *Vodafone* have started proving mobile cloud computing services. *Alibaba* has launched its cloud based mobile operating system in July 2011 [5].

According to the authors of [31], apart from promoting mobility and portability, the advantages of *MCC* lie in a more efficient power consumption; increasing the computational capabilities of mobile applications in terms of CPU speed, memory capacity, storage capacity; and improved reliability. In fact, the execution of remote applications can lower the power consumption considerably. The authors of [68] provide a recent discussion of the implications for the creation of energy efficient *MCC*. The usage of *MCC* can increase the computational and storage capacities of mobile applications, and thus allow to execute computationally and memory/storage expensive applications such as data mining applications on mobile devices. For example, according to [44], *MCC* can be used for transcoding for the delivery of videos to mobile devices. According to [31], the reliability of mobile applications can be improved through storage of application data and application state in the cloud.

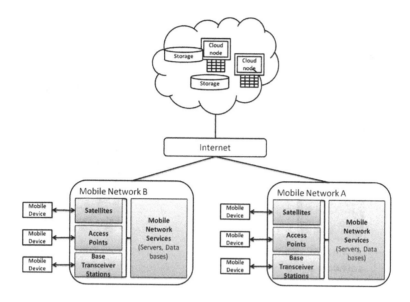

Fig. 2.8 A typical example of a mobile cloud computing architecture [31]

The basic architecture of *MCC* is depicted in Figure 2.8. Essentially mobile devices are connected directly to the so called base stations which can be satellites or more commonly base transceiver stations and access points, which are in turn connected to the network's mobile services, which in turn deliver cloud subscribers requests to the cloud through the Internet. The cloud controllers process the subscribers' requests in order to provide them with the appropriate cloud services.

However, *MCC* services are more of commercial nature for business users rather than sharing and accessing resources on demand. It may actually be useful to implement a grid/cloud infrastructure on the phones themselves and share resources for modest sized analysis tasks, thus the power consumption would be shared as well. For example, for reliability issues it may be useful to cluster application state data across multiple smartphones in the network. This will not consume much resources on any of the participating devices, but it will ensure that the application's state is saved in the event of a device failure (for example through low battery levels). Thus, the application could be resumed at the point of failure rather than starting from the outset of the task. *Pocket Data Mining* suits these needs through the sharing of mobile computational resources. For a more comprehensive survey on *MCC* we refer to [31].

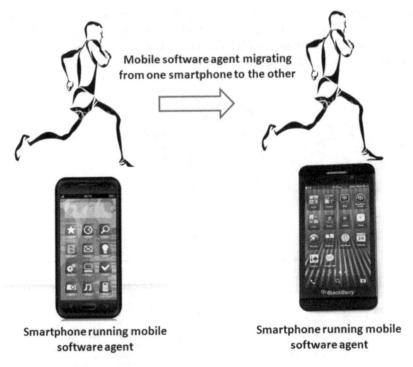

Fig. 2.9 Illustration of the Mobility Feature of Software Agents

2.4 Moblie Agent Technologies

Mobile software agents are computer programs that autonomously and intelligently move from one node to the other to accomplish its task. There are generally two aspects of this definition: (1) the autonomous nature of the software [104]; and (2) its mobility. Figure 2.9 shows an illustration of a mobile software agent hopping from one smartphone off to the other.

The autonomous nature makes it an excellent candidate as a technology that can be used for flexible and on the fly deployment. This is required in our *PDM* framework, as will be detailed in the following chapters of this book.

The mobility feature of the mobile agent technology is an attractive aspect that makes it a competitor for the long-standing Remote Procedure Call (*RPC*). Degrees of mobility vary from weak to strong. The weak mobility of agents implies that internal status of the running agent is not carried out when the agent migrated from one node to the other. On the other hand, strong mobility of agents allows the internal status of the agents to be transferred from one node to the other upon node migration. Only a handful of agent development frameworks provide strong mobility of agents. It is important to note that weak mobility allows the migration of data with the agents. This is exactly what is needed by one kind of agent in the *PDM* framework that will be detailed in the following chapter.

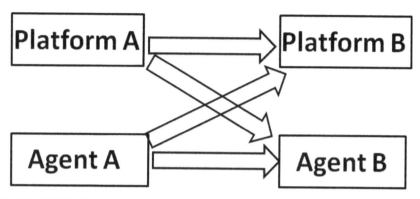

Fig. 2.10 Mobile Software Agent Security Threats

Potential applications and obstacles of this technology have been detailed in [106, 71]. Despite the many important applications the mobile software agents can offer, the security problem is still of concern. There are four main types of security problems that face mobile software agents and threaten the whole paradigm.

- *Agent to Agent*: agents can communicate to each other information that alters the behavior of running the agent's code. Such malicious agents need to be detected and terminated.
- *Agent to Platform*: malicious agents can harm the device they run on, acting as a computer virus.
- *Platform to Agent*: the platform that hosts the agent may terminate it, or change its behavior by altering its code.
- *Platform to Platform*: sending a malicious agent to another platform can give malicious hosts the opportunity to harm other hosts.

Figure 2.10 shows the four types of security threats in the mobile software agent paradigm. Many proposals addressing the above security threats can be found in the literature. Among such techniques come the work done by Page et al [72] addressing the security of a community of mobile software agents in what was termed *Buddy model*. This model can be of use in our project, as a number of mobile software agents work together to realize the *PDM* framework, as detailed in Chapter 3.

The use of mobile agent technology for distributed data mining has been recognised, as an alternative paradigm to the client/server technologies for large databases. A cost model has been developed by Krishnaswamy et al [63], suggesting a hybrid approach to distributed data mining, combining both client/server and mobile agent paradigms. Our choice of mobile agent paradigm in this research project has been due to the fact that our approach follows a peer-to-peer computation mode, and also that centralization of the stream data mining in the mobile computing environment is infeasible. Most importantly, mobile software agents offer the flexibility of deployment of data mining techniques for ad hoc tasks, which is a requirement of the *PDM* process.

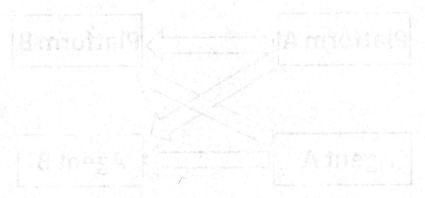

Chapter 3
Pocket Data Mining Framework

This chapter is the core chapter of this monograph. We provide in sufficient details the description of our *Pocket Data Mining (PDM)* framework, including an experimental study that proves the feasibility of *PDM*.

3.1 Introduction

PDM describes the process of performing collaborative data stream mining locally on mobile devices utilizing the flexibility provided by the mobile software agent technology. The process is executed in an ad hoc mobile network of potentially a large number of mobile devices.

Section 3.2 illustrates the basic framework and workflow of *PDM*, whereas Section 3.3 gives practitioners details of the implementation of *PDM* on a desktop computer and highlights a particular implementation of *PDM* using two different classifiers. In the following chapter, details about porting *PDM* to the mobile environment are discussed.

3.2 PDM Architecture

PDM is a generic process to perform mining of streaming data in a mobile environment. The basic architecture of *PDM* is depicted in Figure 3.1. The mobile device that requires a data mining task and utilizes *PDM* to solve this task is referred to as the *task initiator*. *PDM* consists of three generic software agents that can be mobile, and thus are able to move between mobile phones within an ad hoc mobile network [92]. Mobility of software agents in this application is weak mobility; agents move, but do not maintain their internal status of execution. In the following, the three different software agents in *PDM* are discussed.

M.M. Gaber, F. Stahl, and J.B. Gomes, *Pocket Data Mining*, Studies in Big Data 2,
DOI: 10.1007/978-3-319-02711-1_3, © Springer International Publishing Switzerland 2014

3.2.1   Mobile Agent Miners

(Mobile) Agent Miners *(AM)* are distributed over the ad hoc network. They may be static agents used and owned by the user of the mobile devices. Alternatively they may be mobile agents remotely deployed by the user of a different mobile device. They implement the basic stream mining algorithms. However, they could also implement batch learning algorithms if required by the application.

Mobility of this type of agents allows for flexibility of deployment of the data mining algorithms. Additionally it can serve as a space saving strategy when the mobile devices that would form the ad hoc network are known in advance. *AM*s can be distributed on such devices such that each *AM* resides at one device only. When the data mining process starts, dynamic deployment with the help of the mobility feature of the software agents can be used. However, if the devices are not known in advance, each device will be checked for data mining algorithms hosted. This will be achieved using the second type of software agents in *PDM* highlighted in the next section.

3.2.2   Mobile Agent Resource Discovers

Mobile Agent Resource Discoverers *(MRD)* are mobile agents that are used to roam the network in order to discover for the data mining task relevant data sources, sensors, *AM*s and mobile devices that fulfill the computational requirements. They can be used to derive a schedule for the *Mobile Agent Decision Makers* described in the section below.

*MRD*s play an important role in the dynamic deployment of *AM*s. They serve as a landscape discovery of the already formed mobile ad hoc network. Basically decisions to do with which *AM* should run on any specific mobile devices in the network are made by the *MRD*s. Such a decision is done matching the mobile device capability and the data sources the mobile device has access to with the appropriate *AM* according to the data mining task stated by the task initiator.

3.2.3   Mobile Agent Decision Makers

Mobile Agent Decision Makers *(MADM)* can roam through the mobile devices that run *AM*s and consult the *AM*s in order to retrieve information or partial results for the data mining task. The *MADM*s can use the schedule derived by the *MRD*s. *MRD*s play the role as the result aggregating agent. This task varies from one data mining process to the other. For example, if a classification process is needed, *MADM*s can collect the class labels predicted by each of the classifiers *(AMs)*. However, if a clustering process is to run, the *MADM*s will be tasked with the collection of statistics about the clustering results produced by each *AM* for result fusion, producing a global clustering model.

Knowledge integration is a well-studied topic in distributed data mining. *MADM* is basically performing this task on the move. This move also helps the *MADM* to terminate its itinerary early, if a definite decision has been reached. For example, if *n* devices are to be visited by the *MADM*, each of which has the same weight in providing the class label in a classification process, it would be sufficient if $\frac{n}{2} + 1$ nodes are visited with all giving the same label.

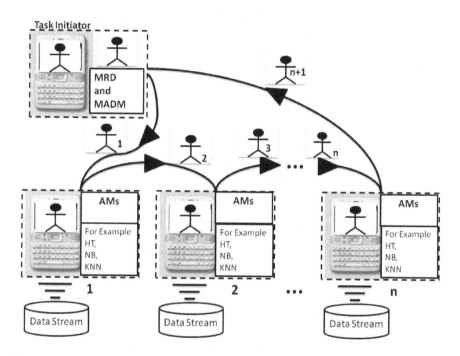

Fig. 3.1 The PDM Architecture

3.2.4   PDM Workflow

Having discussed the three software agents in *PDM*, the process of performing a data mining task utilizing *PDM* is described. Algorithm 1 describes the basic data mining workflow that *PDM* employs collaboratively. The task initiator forms an ad hoc network of participating mobile devices within reach. Next the task initiator starts the *MRD* agent which will roam the network searching for data sources that are relevant for the data mining task, and for mobile devices that fulfill the computational requirements (battery life, memory capacity, processing power, etc.). While the *MRD* is collecting this information, it will decide on the best combination of techniques to perform the data mining task. On its return to the task initiator, the *MRD* will decide which *AM*s need to be deployed to remote mobile devices. There might be restrictions, some mobile phone owners may not allow 'alien' *AM*s to be deployed, for example, they may have limited computational capacity such as battery life, their data might be confidential, etc. Concerning the confidentiality issues the owner of the mobile device may still allow its own *AM*s to be consulted as s/he will have control over its own *AM*s, and thus about which information they release to alien *MRD* and *MADM* agents. The *AM*s are executed concurrently as indicated by the parallel for loop (parFor) in Algorithm 1. Finally the task initiator starts the MADM agent with the schedule provided by the MRD agent. The MADM agent visits the in the schedule listed AMs and will use their model in order to gain information for their decision making process. Finally on the return to the task initiator the *MADM* agent will make a collective decision based on the information gathered from the *AM*s distributed in the network.

Task Initiator: Form an ad hoc network of mobile phones;
Task Initiator: start *MRD* agent;
MRD: Discover data sources, computational resources and techniques;
MRD: Decide on the best combination of techniques to perform the task;
MRD: Decide on the choice of stationary AMs and deploy mobile *AM*s;
Task Initiator: Start *MADM* agent with schedule provided by the *MRD*;
parFor $i = 1$ **to** $i = number\ of$ AMs **do**
    AM_i: starting mining streaming data until the model is used by the MADM agent.
end parFor

Algorithm 1. PDM's collaborative data mining workflow

The current implementation of *PDM* offers *AM*s that implement classifiers for data streams, in particular the *Hoeffding trees* and the *Naive Bayes* classifiers which will be described later in this chapter. It is assumed that the *AM*s are subscribed to the same data stream, however, potentially to different parts of it. For example, in the context of the stock market application, such as *MobiMine* briefly highlighted in Section 2.1, the mobile device may only have data about shares available in which the user of the mobile device is interested in, or more general the mobile device may only be subscribed to specific features of the data stream. A broker may train his/her own *AM* classifier on the data s/he has subscribed to, this could be for example by updating a model which is based on classes 'buy', 'sell', 'do not sell' and 'undecided' whenever s/he makes a new transaction. He may also use the current model to support his/her decisions to 'buy' or 'sell' a share. However, if the broker is now interested in buying a new share s/he has not much experience with, thus s/he may be interested in what decisions other brokers are likely to make in the same situation. Other brokers may not want to disclose their actual transactions but may share their local *AM* or even allow alien *AM*s to be deployed and for this the brokers can use *PDM*. With the current version of *PDM* the data mining workflow outlined in Algorithm 1 may look like the following, where the mobile device of the broker interested in investing in a new share is the task initiator. In the steps below and elsewhere in the book, if we refer to a *PDM* agent hopping, we mean that the agent stops its execution, and is transferred by *PDM* to a different mobile device and resumes its execution on this device. Also in the steps below, it is assumed that the ad hoc network is already established.

1. Task Initiator sends an *MRD* agent in order to discover mobile devices of brokers that have subscribed to relevant stock market data, i.e., data about the shares the broker is interested in.
2. The *MRD* agent hops from one mobile device to another, and if it finds a device subscribed to relevant data, it memorizes the device, and also if there are any useful *AM*s already available. If there are no useful agents, it will memorize if the device allows alien agents to be deployed.
3. The *MRD* agent returns to the task initiator. From there, the *MRD* agent will remotely deploy relevant *AM*s to mobile devices in its list that have relevant data but no relevant *AM*s, however, allow alien *AM*s to be deployed remotely.
4. Once all *AM*s are deployed, the *MRD* agent composes a schedule of all relevant classifier *AM*s subscribed to relevant data and passes it on to the *MADM* agent.
5. The *MADM* agent loads the data about the new shares the broker is interested in and starts hopping to each *AM* in the schedule.
6. On each *AM*, the *MADM* agent hands over the 'shares data' to the *AM* and asks to classify it for example with class 'buy', 'do not buy', 'sell', 'do not sell' or 'undecided. The *MADM* may also retrieve some estimate how reliable the *AM*s thinks its classification is, for example its local classification accuracy.
7. Once the *MADM* returns to the task initiator it may employ a majority voting on the collected classifications from each *AM* or a weighted majority voting incorporating the *AM*s local accuracy (we will call this the *AM*s weight). The outcome of the (weighted) majority voting is used as recommendation for the broker to the investment in the new share.

We used this stock market scenario in this chapter for the illustration of the role of each agent in the *PDM* process. This scenario will be re-visited in Chapter 7 when discussing the potential application of *PDM*.

3.3   PDM Implementation

PDM in its current version offers two *AM*s for classification tasks on data streams. One of the *AM*s implements the *Hoeffding Tree* classifier [32] and one that implements the Naive Bayes classifier. The *AM* that employs the *Hoeffding Tree* classifier uses the *Hoeffding Tree* implementation from the **M**assive **O**nline **A**nalysis (MOA) tool [19].

Hoeffding Tree classifiers have been designed for high speed data streams. It is based on using the *Hoeffding bound* to determine with an error tolerance (ε) and probability $(1 - \delta)$ whether the number of data records/instances seen so far (n) in the data stream is sufficient for finding the attribute to split on at any branch in the decision tree. The *Hoeffding bound* is calculated as follows:

$$\varepsilon = \sqrt{\frac{R^2 \ln(1/\delta)}{2n}}$$

In the context of building the *Hoeffding Tree* incorporating information gain, the range R in the above formula equals $\log_2 L$, where L is the number of class labels in the dataset. It is worth noting that R changes when using a different criteria than information gain for splitting in the decision tree. If the difference in the information gain between the two highest attributes is greater than ε, the algorithm decides that it is statistically guaranteed to use the attribute with the highest information gain. Otherwise, more data records in the stream are observed, until the *Hoeffding bound* is satisfied.

The *Naive Bayes* classifier has been originally developed for batch learning, however, its incremental nature makes it also applicable to data streams. Similar to *Hoeffding Tree*, the *AM* employing the *Naive Bayes* classifier uses the *Naive Bayes* implementation from the *MOA* tool [19]. *Naive Bayes* is based on the *Bayes Theorem* [64] which states that if C is an event of interest and $P(C)$ is the probability that event C occurs, and $P(C|X)$ is the conditional probability that event C occurs under the premise that X occurs then:

$$P(C|X) = \frac{P(X|C)P(C)}{P(X)}$$

The Naive Bayes algorithm uses the *Bayes Theorem* to assign to a data instance to the class it belongs to with the highest probability.

3.4   Case Studies of Distributed Classification for Pocket Data Mining

Three different configurations of *PDM* have been thoroughly tested. One *PDM* configuration is solely based on *Hoeffding Tree AMs*, the second configuration of *PDM* is solely based on *Naive Bayes AMs*, and the third configuration is a mixtures of both *Hoeffding Tree* and *Naive Bayes AMs*. Section 3.4.1 outlines the general experimental setup, Section 3.4.2 outlines the experimental results obtained using only *Hoeffding Tree AMs*, Section 3.4.3 outlines the experimental results obtained using only *Naive Bayes AMs*, and Section 3.4.4 outlines the experimental results obtained using a mix of *Hoeffding Tree* and *Naive Bayes AMs*.

3.4.1   Experimental Setup

As aforementioned, the two classifier *AMs* use the *Hoeffding Tree* and *Naive Bayes* implementations from the *MOA* toolkit [19]. *PDM* is also built on the well known **J**ava **A**gent **D**evelopment **E**nvironment *(JADE)* [16]. *JADE* agents are hosted and executed in *JADE containers* that can be run on the mobile devices and PCs. *JADE* agents can move between different *JADE containers*, and thus between different mobile devices and PCs. As *JADE* agents can be developed on PCs and run on both PCs and mobile phones, it is possible to develop and evaluate *PDM* on a *Local Area Network (LAN)* of PCs. The used *LAN* consists of 9 workstations with different software and hardware specifications and is connected with a *CISCO* Systems switch of the catalyst 2950 series. In the configurations of *PDM*, 8 machines were either running one *Hoeffding Tree* or one *Naive Bayes AM* each. The 9th machine was used as the task initiator, however, any of the 8 machines with *AMs* could have been used as task initiator as well. The task initiator starts the *MADM* in order to collect classification results from the *AMs*.

Table 3.1 Evaluation Datasets

Test Number	Dataset	Number of Attributes	Number of Instances
1	kn-vs-kr	36	1988
2	spambase	57	1999
3	waveform-500	40	1998
4	mushroom	22	1978
5	infobiotics 1	20	≈ 200000
6	infobiotics 2	30	≈ 200000

The data streams for *PDM* have been simulated using the datasets described in Table 3.1. The datasets have been labeled with test 1 to 6 for simplicity when referring experiments to a particular data stream. The data for test 1, 2, 3 and 4 have been retrieved from the *UCI* data repository [21] and datasets 5 and 6 have been taken from the *Infobiotics* benchmark data repository [11]. All datasets are stored in the *Attribute-Relation File Format (ARFF)*[1] and the data stream is simulated by taking a random data instance from the *.arff* file and feeding it to the *AM*. Instances may be selected more than once for training purposes, making sampling with replacement.

As aforementioned in this chapter, each *AM* may be subscribed to only a subset of the total feature space of a data stream, we call this a vertically partitioned data stream. For example, a stock market broker may only subscribe to data about companies s/he is interested in investing in, or a police officer may only access data s/he has clearance for. Even if a user of a mobile device may have access to the full data, the owner of the device may not want or be able to subscribe to unnecessary features for computational reasons, such as bandwidth consumption, the fact that

---

[1] http://www.cs.waikato.ac.nz/ml/weka/arff.html

the more data is processed by *AM*s, the more power will be consumed, the processing time of the data stream is longer the more features are streamed in and need to be processed, or higher subscription fees may be imposed. Although the current subscription may be insufficient for classifying new data instances, the task initiator can send an *MADM* with the unclassified data instances. This *MADM* visits and consults all relevant *AM*s that belong to different owners that may have subscribed to different features that are possibly more relevant for the classification task.

The *MADM* collects predictions from each *AM* for each unclassified data instance and the estimated 'weight' (accuracy) of the *AM*, which it uses to decide on the final classification. In the *PDM* framework, each *AM* treats a streamed labeled instance either as training or as test instance with a certain probability which is set by the owner of the *AM*. The default probability used in the current setup is 20% for the selection as a test and 80% for the selection as a training instance. Each training instance is put back into the stream and may be selected again as training instance, this allows to simulate endless data streams with reoccurring patterns. The test instances are used to calculate the 'weight' of the *AM*. The *AM* also takes concept drifts into account when it calculates its 'weight' by defining a maximum number of test instances to be used. For example, if the number of test instances is 20 and there are already 20 test instances selected, then the *AM* replaces the oldest test instance by the newly incoming test instance and recalculates the 'weight' using the 20 test instances. A more advanced technique addressing the concept drift issue with a formal definition of the problem is discussed in Chapter 5.

After the *MADM* finished consulting all *AM*s in its schedule, it returns to the task initiator and uses the local predictions from each *AM* and the *AM*s' weights in order to derive a final classification using a 'weighted majority voting'. For example, for the classification of one data instance, if there are three *AM*s: *AM1*, *AM2* and *AM3*. *AM1* predicts class A and has a weight of 0.57, *AM2* also predicts class *A* and has a weight of 0.2, and *AM3* predicts class *B* and has a weight of 0.85. The *MADM*'s 'weighted' prediction for class *A* is $0.57A + 0.2A = 0.77A$ and for class *B* $0.85B = 0.85B$. Thus the *MADM* yielded the highest weighted vote for classification *B* and will label the concerning instance with class *B*.

The user of *PDM* can specify which features its *AM* shall subscribe to, however, in reality we may not know the particular subscription. Thus, in the experimental setup, each *AM* subscribes to a random subset of the feature space. More particularly, each *AM* is given access to 20%, 30% or 40% of the total feature space.

Before discussing the results we achieved running *PDM* with different configurations, it is important to define some concepts. The terminology that is used in Sections 3.4.2, 3.4.3 and 3.4.4 is explained in the following.

- The **weight** refers to the local accuracy of the *AM* calculated using randomly drawn test instances from the local data stream.
- **MADM's accuracy** or **PDM's accuracy** is the accuracy achieved by the *MADM* using the test dataset classified by 'weighted majority voting' by the *MADM*.
- **local accuracy** is not to be confused with the weight. The local accuracy is the actual accuracy that a particular *AM* achieved on classifying the *MADM*'s test data. This accuracy is only calculated for evaluation purposes, it would not be

calculated in the real application as the real classifications of the *MADM*'s test set would be unknown.

- the **average local accuracy** is calculated by averaging the local accuracies of all *AM*s. The average accuracy is used to show if the 'weighted majority voting' performs better than simply taking a majority vote.

3.4.2   Case Study of PDM Using Hoeffding Trees

The datasets listed in Table 3.1 are batch files. Using batch files allows us to induce classifiers using batch learning algorithms, and thus to compare *PDM*'s classification accuracy to the ideal case of executing batch learning algorithms on the whole datasets using all attributes. In particular, the *C4.5* [81] and *Naive Bayes* batch learning algorithms have been used from the *WEKA* workbench [50]. The choice of *C4.5* is based on its wide acceptance and use; and to the fact that the *Hoeffding Tree* algorithm is based on C4.5. The choice of *Naive Bayes* is based on the fact that it is naturally incremental, computationally efficient, and also widely accepted.

In general it is expected that the more features the *AM*s have available the more likely it is that they achieve a high classification accuracy, and thus the more likely it is that the *MADM* achieves a high classification accuracy as well. Yet some features may be highly predictive and others may not be predictive and even introduce noise. Thus in some cases having more features available may decrease the *AM*s, and in turn the *PDM*'s accuracy. 70% of the data instances from each dataset in Table 3.1 have been used to simulate the local data stream and the remaining 30% have been used as test data in order to evaluate *PDM*'s and respectively *MADM*'s accuracy. All experiments outlined in this chapter have been conducted 5 times, the average local accuracy of each *AM* has been calculated and recorded as well as *PDM*'s or respectively *MADM*'s average accuracy.

Figure 3.2 shows *PDM*'s average classification accuracy plotted versus the number of *AM*s visited. The experiments have been conducted for configurations where all *AM*s either subscribe to 20%, 30% or 40% of the features of the data stream. The features each *AM* subscribes to are selected randomly, thus some *AM*s may have subsets of their features in common and some not. That two or more *AM*s have features in common is a realistic assumption, for example, for the stock market broker application, two brokers may be interested in the 'Compaq' share, but only one of them may be interested in the 'Hewlett-Packard' share, and the other in the 'Microsoft' share.

The largest difference between *Naive Bayes*'s accuracy and that of *C4.5* is for test 2 where *Naive Bayes*'s accuracy is 80% and *C4.5* reach 91%, otherwise both batch learning algorithms achieve similar accuracies. Concerning *PDM*'s accuracy based on *Hoeffding Trees*, it can be seen that *PDM* generally achieves accuracies above 50% for all datasets. In general, *PDM* configurations with *AM*s using just 20% of the feature space perform much worse than configurations with 30% or 40% which can be explained by the fact that predictive features are more likely not to be

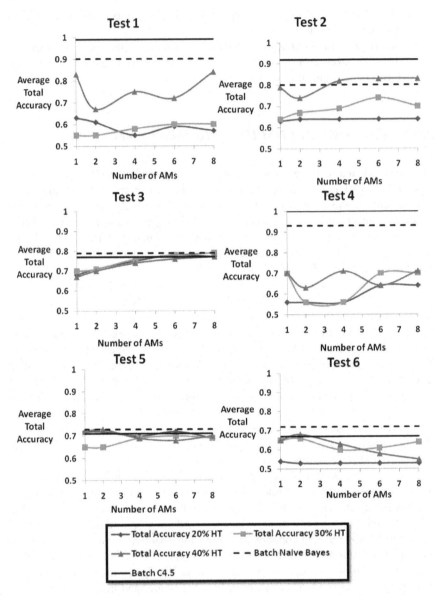

Fig. 3.2 PDM's average classification Accuracy based on Hoeffding Trees

selected. In some cases, for example, for test 2 it seems that configurations of *PDM* with 30% achieve a higher accuracy than configurations with 40% which can be due to the fact that with subscribing to 40% of the features it is also more likely that non predictive features that introduce noise are selected compared with subscribing to 30%. In general, *PDM* achieves accuracies close to the batch learning algorithms

C4.5 and *Naive Bayes*, notably in tests 3 and 5, but also for the remaining tests *PDM* achieves close accuracies to those of *Naive Bayes* and *C4.5*. It can also be observed that *PDM* based on *Hoeffding Trees* achieves acceptable classification accuracy in most cases.

Varying the number of *AM*s generally is dependent on the dataset used. Highly correlated attributes in one dataset would only need a small number of *AM*s and vice versa.

Figure 3.3 compares *PDM*'s accuracy (achieved by the *MADM* through 'weighted majority voting') with the average local accuracy of all *AM*s versus the number of *AM*s visited. Each row of graphs corresponds to one of the tests in Table 3.1 and each column of graphs corresponds to a percentage of features the *AM*s are subscribed to. The lighter line in the graphs is the accuracy of *PDM* and the darker line is the average local accuracy of all *AM*s. *PDM*'s accuracy is in most cases higher or even better than the average local accuracy, hence the *MADM*'s 'weighted majority voting' achieves a better result compared with simply taking the average of the predictions from all *AM*s.

3.4.3   Case Study of PDM Using Naive Bayes

A further configuration of *PDM* solely based on *Naive Bayes AM*s has been evaluated the same way as *PDM* solely based on *Hoeffding Trees* has been. *PDM* solely based on *Naive Bayes* is expected to produce similar results compared with *PDM* solely based on *Hoeffding Trees* evaluated in Section 3.4.2.

Figure 3.4 presents the data obtained of *PDM* solely based on *Naive Bayes* the same way as Figure 3.2 does for *PDM* solely based on *Hoeffding Trees*. Similarly the experiments have been conducted for configurations where all *AM*s either subscribe to 20%, 30% or 40% of the features of the data stream. The features each *AM* subscribes to are selected randomly, thus some *AM*s may have subsets of their features in common and some not. Concerning *PDM*'s accuracy based on *Hoeffding Trees*, it can be seen that *PDM* generally achieves accuracies above 50% for all datasets. Similarly compared with Figure 3.2, *PDM* configurations with *AM*s using just 20% of the feature space generally perform much worse than configurations with 30% or 40% which can be explained by the fact that predictive features are more likely not to be selected. Yet in some cases, for example for test 2 it seems that configurations of *PDM* with 30% achieve a higher accuracy than configurations with 40% which can be due to the fact that with subscribing to 40% of the features it is also more likely that non predictive features that introduce noise are selected compared with subscribing to 30%.

In general, *PDM* achieves accuracies close to the batch learning algorithms *C4.5* and *Naive Bayes*, notably in tests 3, 4, 5 and 6. However, also for the remaining tests *PDM* achieves close accuracies to those of *Naive Bayes* and *C4.5*. More generally, *PDM* based on *Naive Bayes* achieves acceptable classification accuracy in most cases.

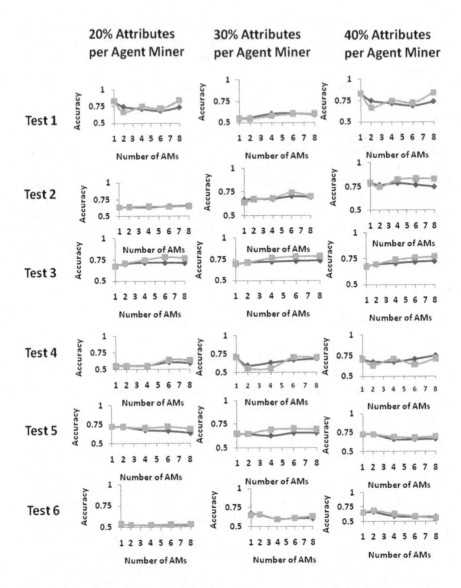

Fig. 3.3 PDM's average classification accuraciy versus the average local accuracy of the AMs with Hoeffding Trees

Similar to the previous set of experiments, varying the number of *AM*s generally is dependent on the dataset used. Highly correlated attributes in one dataset would only need a small number of *AM*s and vice versa.

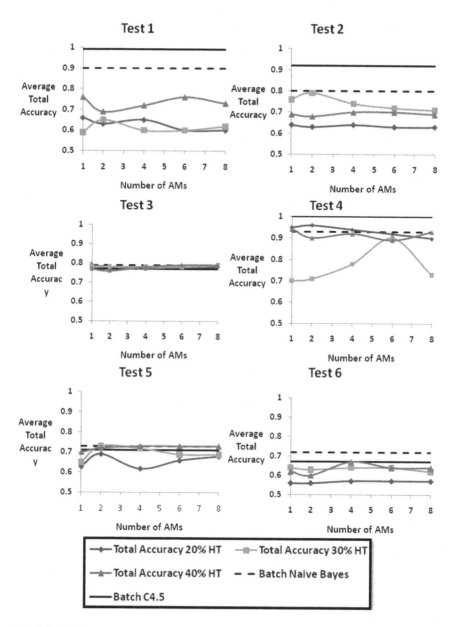

Fig. 3.4 PDM's average classification Accuracy based on Naive Bayes

Figure 3.5 analogous to Figure 3.3 opposes *PDM*'s accuracy (achieved by the *MADM* through 'weighted majority voting') and the average local accuracy of all *AM*s versus the number of *AM*s visited. Each row of graphs corresponds to one of the tests in Table 3.1 and each column of graphs corresponds to a percentage of features

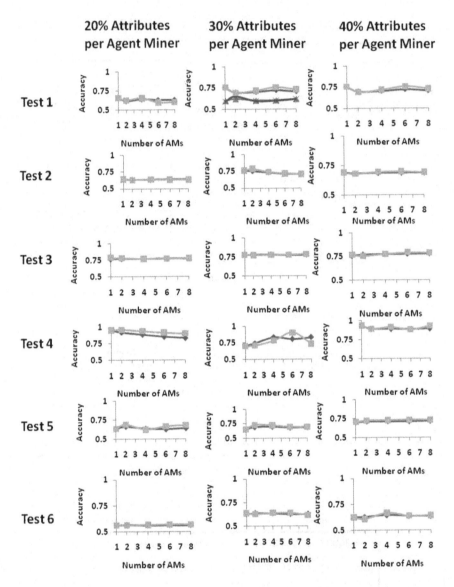

Fig. 3.5 PDM's average classification accuracy versus the average local accuracy of the AMs with Naive Bayes

the *AM*s are subscribed to. The lighter line in the graphs is the accuracy of *PDM* and the darker line is the average local accuracy of all *AM*s. Similar to the *Hoeffding Tree* results *PDM*'s accuracy is in most cases higher or even better than the average local accuracy, hence the *MADM*'s 'weighted majority voting' either achieves a better result than simply taking the average of the predictions from all *AM*s.

3.4.4   Case Study of PDM Using a Mixture of Hoeffding Trees and Naive Bayes

Figure 3.6 highlights the accuracies of the two *PDM* configurations solely based on *Hoeffding Trees* and solely based on *Naive Bayes* for different numbers of visited *AM*s. The bars in the figure are in the following order from left to right: Total accuracy of PDM with *Hoeffding Trees* with 20% attributes; total accuracy of *PDM* with *Hoeffding Trees* with 30% attributes; total accuracy of *PDM* with *Hoeffding Trees* with 40% attributes; total accuracy of *PDM* with Naive Bayes with 20% attributes; total accuracy of *PDM* with *Naive Bayes* with 30% attributes; total accuracy of *PDM* with *Naive Bayes* with 40% attributes; accuracy for batch learning of *Naive Bayes* with all attributes; and finally accuracy for batch learning of *C4.5* with all attributes.

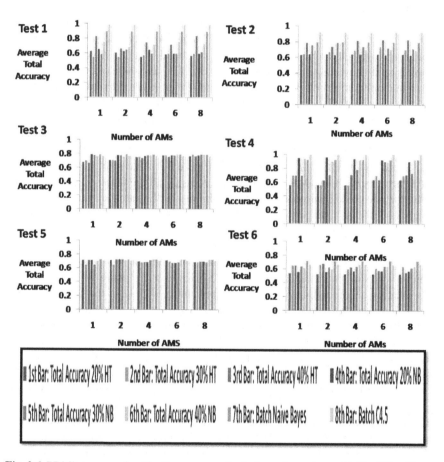

Fig. 3.6 PDM's average classification accuracy for both configurations with Hoeffding Trees and Naive Bayes

On tests 3 and 5, both configurations of *PDM* achieve almost equal classification accuracy. Also for tests 1, 2, 4 and 6, the classification accuracies of *PDM* are very close for both configurations and there does not seem to be a bias towards one of the classifiers used. Hence a heterogeneous configurations of *PDM* with a mixture of both classifiers would be expected to achieve a similar performance than a homogeneous configuration solely based on *Hoeffding Trees* or *Naive Bayes*. Such a heterogeneous setup of *AM* classifiers would also be a more realistic setup as owners of mobile devices may use their individual classification techniques tailored for the data they subscribed to.

In order to show that a heterogeneous setup of *AM* classifiers achieves a similar accuracy to a homogeneous solely based on *Hoeffding Trees* or *Naive Bayes*, we have evaluated configurations of *PDM* that use both classifiers. All experiments have been conducted five times and the average of the achieved accuracy by the *MADM* has been calculated.

Figure 3.7 highlights experiments conducted with different heterogeneous setups of *PDM* for all 6 datasets listed in Table 3.1. The average accuracy of the *MADM* is plotted against the combination of algorithms embedded in the deployed *AM*s. The graph is split showing *AM*s working with 20%, 30% and 40% of the features. All possible combinations of *Hoeffding Trees* and *Naive Bayes AM*s have been evaluated. The horizontal labels in Figure 3.7 are read the following way. *HT* stands for *Hoeffding Tree* and *NB* for *Naive Bayes*, the number before *HT* and *NB* is the number of *Naive Bayes* or *Hoeffding Tree* classifiers visited respectively. For example label '3HT/5NB' means that 3 *Hoeffding Tree AM*s and 5 *Naive Bayes AM*s have been visited by the *MADM*. Also plotted in Figure 3.7 is the result the batch learning algorithms *C4.5* and *Naive Bayes* achieved using all the features. The achieved accuracies are close compared with those achieved by the batch learning algorithms which have the advantage over *PDM* of having all the features available, which would again not be the case in a realistic scenario where subscribers of a data stream limit their subscription only to properties they are particularly interested in.

In Figure 3.7 for test 1, it can be seen that using 20% or 30% features seems to achieve very similar classification accuracies. However, for using 40% features the classification accuracy improves considerably and gets close to the batch learning accuracies which use all features, also for using 40% of the features it can be seen that configurations with more *Hoeffding Tree AM*s perform slightly better than configurations with more *Naive Bayes AM*s. For test 2, it can be seen that using 30% instances already improves the classification accuracy and there is a tendency for the usage of 30% and 40% instances that configurations with more *Hoeffding Tree AM*s perform slightly better, in particular configurations 6HT/2NB and 7HT/1NB seem to achieve high accuracies between 80% and 90%. Regarding test 3 all percentages of features used achieve a very similar and very good classification accuracies that can well compete with the accuracies achieved by the batch learning algorithms on all features. Also configuration wise it seems that using configurations with more *Naive Bayes* than *Hoeffding Trees* seem to perform better, however, also a configuration solely based on *Hoeffding Trees* achieves very good classification accuracies close to batch learning algorithms. In test 4 the tendency seems that using more

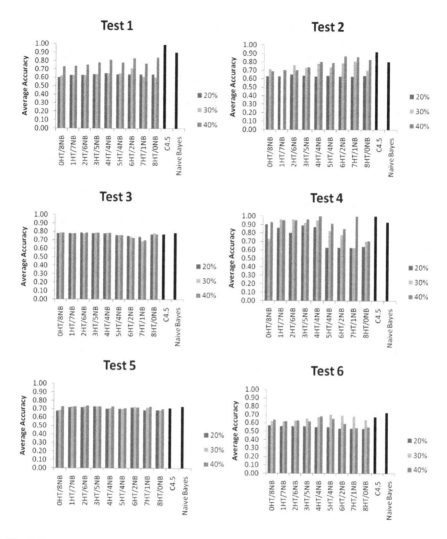

Fig. 3.7 Average classification accuracy for a heterogeneous setup of classifier AMs in PDM

Naive Bayes classifiers achieve a higher accuracy than using more *Hoeffding Tree* classifiers. In test 4 *Naive Bayes* seems to work better on configurations with less features subscribed to compared with *Hoeffding Trees*. There is no noticeable tendency for test 5, all configurations and percentages of features seem to achieve a high accuracy very close to the one observed for the batch learning algorithms. On test 6 any configuration and even batch learning algorithms do not achieve a good classification accuracy. This suggests that the dataset for test 6 is not very well suited for classification in its current form. Also there is no particular tendency detectable for test 6.

Figure 3.7 also displays the data from homogeneous configurations of *PDM* solely based on *Hoeffding Trees*, which is labeled as configuration 8HT/0NB and solely based on *Naive Bayes*, which is labeled 0HT/8NB. The results clearly show that heterogeneous configurations of *PDM* achieve very similar accuracies compared with homogeneous configurations of *PDM*. Also earlier in this section we stated that a heterogeneous setup of *AM* classifiers would also be a more realistic setup as owners of mobile devices may use their individual classification techniques. Furthermore, *PDM* may well benefit from using individual *AM*s from different owners as they are likely to be optimized on the local subscription of the data stream.

3.5   Concluding Remarks

This chapter presented the core of our *PDM* framework, providing its feasibility from the predictive accuracy perspective. We gave detailed discussions of a thorough experimental study varying the mining algorithms used, the size of the random subspace, and heterogeneity of the used algorithms. The study clearly showed the potential of the *PDM* framework.

However, realizing *PDM* as a mobile technology can not be evidenced without running *PDM* in a real mobile network, running mobile software agents. The following chapter gives practitioners the details of porting *PDM* from the desktop to the smartphone environment.

Chapter 4
Implementation of Pocket Data Mining

This chapter describes the main design decisions made to implement the first proto-type of *PDM* in its intended environment, mobile devices. This is followed by a high level walk through on setting up a collaborative classification using the *PDM* imple-mentation. Finally, some limitations of the current *PDM* prototype implementation are discussed.

The chapter serves as a reference point for practitioners interested in develop-ing *PDM*-based applications. It can also help those interested in utilizing mobile software agents in smartphone applications.

4.1 Choice of Software Packages and Components in PDM

The *PDM* framework requires the following software components: (1) an operat-ing system for the mobile devices, (2) a mobile agent platform to host the *PDM* agents outlined in Chapter 3, and (3) the data mining software/libraries used to add functionality to *PDM*.

4.1.1 The Mobile Operating System

Currently the smartphone industry is undergoing a major shift, where new operating systems (*OS*) have emerged, namely *iOS* and *Android* , which are currently pushing their rapid development. Some years ago the usage of a smartphone would not have seemed appropriate for an ordinary user, as it offered some complex functionality that would barely be useful and would have imposed a steep learning curve for the user. The introduction of these operating systems changed the smartphone's pop-ularity. Thanks to an improved *Graphical User Interface (GUI)*, ease of usage, af-fordable prices, better hardware and the existence of useful applications, contributed to the success of smartphones for the every day usage. The other type of phones,

M.M. Gaber, F. Stahl, and J.B. Gomes, *Pocket Data Mining*, Studies in Big Data 2,   41
DOI: 10.1007/978-3-319-02711-1_4, © Springer International Publishing Switzerland 2014

commonly known as *feature phones*, have less computational power and offer less functionality with the reasoning that a mobile phone is to be used mainly for calls and messages and probably a few extras. The main advantage of this type of phones is that they are cheaper than smartphones and offer better battery performance, providing more usage time between charges. It is very common to find a customised *OS* on these mobile phones as it has to be tuned for the specific hardware for better performance. This in turn makes it difficult to write programs that work on different mobile phone platforms if they are not running the same *OS*. Thus these phones usually support *Java ME (Micro Edition)* in order to run third party programs. The biggest problem with *Java ME*, is that it was designed for computational constrained devices, so it attempts to support only the bare minimum functionality while avoiding more complex tasks.

The idea for *PDM* is to have a multi-platform implementation, capable of running seamlessly across different mobile devices. But nevertheless, a mobile *OS* has to be chosen as the first prototype. For testing purposes, low costs are preferred, plus at least an *OS* with a decent market share and enough power to support data mining.

In this case *Android* seems to be the best choice, as it is open source, offers a free software development kit and according to a *Nielsen* study shown in Figure 4.1, it has the biggest acquisition share as of September 2011 in the United Kingdom. Nonetheless, other *OS*s should not be neglected. *Android* has simply been chosen as the first implementation target.

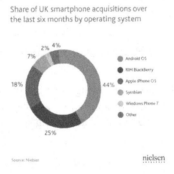

Fig. 4.1 Smartphone acquisition rate in the UK in 2011[1]

4.1.2   The Mobile-Agent Platform

Mobile agent technology has been around for over a decade, which has allowed it to evolve and to offer different approaches. During this time some frameworks have been developed to ease the use of this technology and to ensure the interoperability

---

[1] http://www.nielsen.com/uk/en/insights/press-room/2011-news/
more-uk-consumers-choose-android-smartphones--
but-many-still-cov.html

of the agents. Examples of such frameworks are *Grasshopper*, *Java Agent Development Environment (JADE)*, *Java Agent Services (JAS)*, *Lightweight Extensible Agent Platform (LEAP)*, *RETSINA*, *ZEUS* and *Mole*[2,3]. To avoid spending time comparing the different frameworks, one must be chosen based on the implementation requirements. Basically the implementation needs a system capable of running several agents across distributed hosts and the ability to move an agent between the hosts (for example for *PDM*'s *MADM* agent). Also, it is advisable to use a framework that conforms to a standard like *FIPA (Foundation for Intelligent Physical Agents)*, which is supported by the *IEEE*[4], as this offers possibilities of mobile agents interoperability between other systems. The well known *JADE* framework just fits these requirements. One of its other strong points is that it is open source, and written in *Java*.

4.1.3   Data Mining Libraries

For the implementation of functionality (i.e. *PDM*'s *AM* agents) we used the aforementioned in Chapter 3, the *Massive Online Analysis (MOA)* [19] which is based on the *Java* programming language and some *WEKA* [103] libraries. The reasoning behind this choice is that *MOA* is a popular open source data mining tool that offers a wide range of data stream mining algorithms and is also easy to extend. Figure 4.2 shows *MOA* running a data stream classification technique.

4.2   Background of Software Components and Programming Language Used in PDM

In this section, we shall provide a brief description of the software components and programming language used in developing *PDM*.

4.2.1   The Java Programming Language

Java is an imperative, object-oriented programming language. It was designed with portability in mind. To achieve this goal, instead of compiling to native code, it uses special instructions known as the *Java bytecode*. Instead of generating an executable file (for example, a *.exe* file for Windows), it generates *.class* files, which can be

---

[2] http://www.fipa.org/resources/livesystems.html

[3] http://www.dia.fi.upm.es/~phernan/AgentesInteligentes/
referencias/bellifemine01.pdf

[4] http://www.ieee.org/index.html

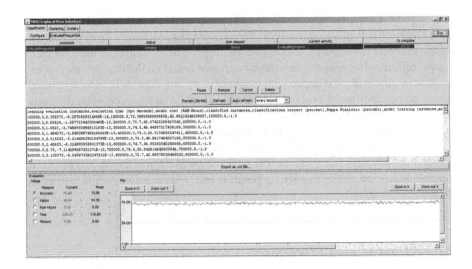

Fig. 4.2 MOA Graphical User Interface(GUI)

further grouped as a single *.jar* file. Thus the compiled program will not be able to run directly on one's computer *OS*, but at the same time is not constrained to a single instruction set architecture (i.e. operating system). In order for *Java* programs to run, some sort of middleware is required. This is where the *Java* platform comes into play. It offers a set of libraries used to perform common tasks, and the *Java Virtual Machine* is in charge of translating the special *bytecode* instructions into the *OS* native instructions.

The Java platform has 3 main variants, which basically differ by the libraries offered:

- *Standard Edition*: offers the main functionality of the *Java* environment.
- *Enterprise Edition*: offers the whole functionality found in the *standard edition*, plus some libraries tailored specifically for the enterprise market.
- *Micro Edition*: offers only a subset of the functionality found in the *standard edition* and some additional libraries optimised for resource-constrained devices.

The core *Java* libraries, known collectively as the *Java Class Library*, are open source. These libraries offer a variety of functionality, mostly for common tasks and as a means to interact with the hardware. They can be classified as: base libraries, integration, and user interface libraries[5]

---

[5] http://docs.oracle.com/javase/7/docs/technotes/ guides/index.html

4.2.2   The Android Operating System

Android is an open source operating system optimised for mobile devices. It is currently being developed and maintained by the *Open Handset Alliance*[6].

Android is based on *Linux*, but does not expose its full functionality, as it does not allow the execution of native programs. In order to allow the development of third party applications, *Android* presents its own application framework, which is based on *Java*. As discussed in Section 4.2.1, a virtual machine is required to run the programs. In this case, *Android* offers its own implementation, called the *Dalvik Virtual Machine*, which is optimized to have several instances running simultaneously on a mobile device. This particular virtual machine is not able to run the *.class* files generated by Java, but rather runs Dalvik executable *(.dex)* files, which are devised in such a way that they use as less memory as possible. To generate these files, a developer can use the *dx* tool, contained in the *Android SDK*. This tool basically converts *.class* files into *.dex* files. Also, instead of packing them as a *.jar* file, the files are grouped as an *Android* application package *(.apk)* file, so that *Android* programs can be differentiated from conventional Java programs.

The core libraries available under *Android* provide most of the functionality presented by the base libraries of the *Java* platform *standard edition*. This can be very helpful when porting a *Java* application to *Android*, as most of the code will work exactly the same.

4.2.2.1   Android's Graphical User Interface (GUI)

While Java offers the *Abstract Window Toolkit (AWT)* and *Swing* libraries to deal with the user interface, *Android* has its own framework. *Android*'s framework is designed for smaller screens and varying sizes (also known as fragmentation). The main idea behind this implementation is that the *GUI* can be represented as a hierarchy tree using two types of nodes: the *View*, a *Leaf Node*; and the *ViewGroup*, commonly used as a branch node. The way in which developers can define how components should be arranged inside a *ViewGroup* is through a layout. It could be a linear layout (objects placed one after another, either vertically or horizontally), a table layout, a relative layout or simply a frame layout (in which the objects just stack). These layouts are typically defined in an *XML* file, which already represents a tree data structure, but can be also coded in the program itself. If the *XML* approach is used, then the developer needs to interact with a *GUI* component through code; the method *findViewById* will return a reference to the given object.

---

[6] http://www.openhandsetalliance.com/

4.2.2.2   Android's Activities

The idea behind activities is to break down the actions a user can do on an application. Each activity represents something that the user can do and most of the time is directly tied to the touch screen, which displays for the user the relevant data.

4.2.2.3   Android Services

There might be certain *Android* activities that should be running in the background, i.e., transferring data (to avoid an interruption if the user moves to a different activity or switches the active application). That is where services come to play. Basically a service is the way in which *Android* handles background activities. A service will always run in the main thread of the program, thus if it is doing a heavy process, it might block the whole application. In order to avoid this, threads can be used, as in standard desktop application's programming.

4.2.2.4   Android Security

Mobile phones can handle a lot of personal information and it is important to keep it safe. Also this information can be tied to services (phone calls, *SMS* messages, data usage) that can cost the user money. So it is important to have a security policy and enforce it. First of all, it is necessary to avoid 'privilege scalation', meaning that applications should not be able to affect other applications. *Android* supports multitasking, therefore many applications can be executed at the same time. In order to prevent unauthorized access between applications, each application needs to run in a different *Dalvik Virtual Machine*. It works over a *Linux* kernel, but to keep applications on a sandbox, it runs them on different virtual machines (*Dalvik*)

One of the main differences between *Android* development and classic desktop development is the use of *permissions*. For certain activities *Android* requires that the application specifies that it will be executing them, thus declaring it will need special privileges. By doing this, the user can review the privileges the application is asking for and decide whether or not to install it. If the developer fails to declare these permissions on the application manifest, then the activity will simply fail triggering a 'not enough privilege's error'.

4.2.3   Foundation for Intelligent Physical Agents (FIPA)

In order to keep different agent systems interoperable, a standard was defined. *FIPA* is currently supported by the *IEEE* and basically defines the way in which agents should communicate with each other. By adopting this specification a system can be built that works with other systems regardless of the platform or programming language used, as they share a common communication protocol.

4.2.4 Java Agent Development Environment (JADE)

JADE is a software framework to develop mobile agent applications in compliance with the *FIPA* specifications for interoperable intelligent multi-agent systems. The goal is to simplify development while ensuring standard compliance through a comprehensive set of system services and agents [16].

An important feature that the *JADE* platform provides is the ability to develop distributed systems on several devices (not requiring the devices to have the same operating system). *JADE* even allows to manipulate agents and containers hosted on other computers using a *GUI*. Another important characteristic of this platform is the agent mobility. *JADE* allows migration of agents between containers in order to develop more dynamic and environment-aware systems.

The *JADE* platform was developed by *TILAB* in Milan, Italy[7] and currently is an open source project, which allows us to manipulate the code at will. To date, it has been widely used for conventional desktop computer and laptop environments. However, in this book we explore the possibility of implementing data mining in a collaborative environment developed for mobile devices such as smartphones or tablet computers.

In its continuous effort to consolidate the *JADE* platform, *TILAB* has already started developing versions for mobile devices. As of October 2011, they offer a version for the *Android* Mobile *OS*. Nevertheless, the shortcoming of this official version, is that it only supports one agent per mobile device. They made this decision based on the fact that most phones have limited resources. Although, with the ever increasing capabilities of smartphones, these devices have started to have specifications comparable to that of notebooks, thus this assumption might not hold true for long.

Because of this, the first part of the implementation focuses on the possibility to migrate the functionality of the *JADE* platform to the *Android* operating system. The following is a brief overview of *JADE*.

4.2.4.1 Platforms and Containers

Each *JADE* instance is called container because it may contain several agents. A set of active containers (distributed in several computers) is called a platform. A single special *Main Container* must always be active in a platform and all other containers must register with it as soon as they start. The non-main containers must know the host's address and the port where the main container is allocated, which is where

---

[7] http://jade.tilab.com/

they must register. Later on in this chapter we will discuss how to accomplish this in a user friendly manner.

4.2.4.2   Agent Management System and Directory Facilitator

A main container must contain two special agents that start automatically when this container starts. Those agents are called *Agent Management System (AMS)* and *Directory Facilitator (DF)*.

- The *AMS* provides the names for the created agents and makes sure that there are no duplicated names.
- The *DF* provides a communication service to the agents so they can communicate between each other.

Every agent has a particular goal and every agent performs certain activities that are defined as behaviors. A behavior represents a task that the agent carries out. An agent may have one or more behaviors. A behavior can start out of another behavior.

4.2.4.3   Agent Communication Language Messages

One important feature in *JADE* is communication among agents. Agents communicate by means of messages. A message is the information that one agent sends to one or more other agents. Messages have established *Agent Communication Language (ACL)* formats which are specified by *FIPA*.

In the past few years there have been some efforts to migrate the platform to mobile environments by the developers community. Back in 2008, the first light version of *JADE* for *Android* was released. This version was forked from their previous implementation on *Java ME* devices, thus designed for limited resources and due to its restrictions, it is only possible to create one agent per container at a given time. This turns out to be a disadvantage on platforms where agent migration is required, as is the case of *PDM*. Given this circumstance we found it is more appropriate to focus on adapting the desktop version, instead of trying to get more from the existing *Light JADE* version designed for *Android*.

An important characteristic found while studying the original structure of *JADE* was the use of the *GUI*. Most of the GUI is developed using the library *javax.swing*, which is the *Java* library that replaced the old *Abstract Window Toolkit*. Nevertheless, in the *Android* Operating System it is not possible to migrate such packages, because it has its own libraries for *GUI* development. Fortunately, for the development of standard *JADE* the *GUI* is separated from the actual code.

4.2.5 The MOA Data Stream Mining Software and Library

Probably the most important contribution to the dissemination and exploitation of the advances in data stream mining has to be credited to the aforementioned open source *Massive Online Analysis Tool (MOA)* [19]. The fact that *MOA* offers a selection of data stream mining algorithms and provides an *API* that allows to easily integrate new algorithms, makes *MOA* the ideal data stream mining library to be used by *PDM*'s *AM* agents.

4.2.6 The WEKA Data Mining Software and Library

Weka is open source software developed and maintained by the Machine Learning Group at the University of Waikato [103]. It is a *Java* library designed for data mining and includes several machine learning algorithms. The fact that parts of *MOA*, outlined in Section 4.2.5, use the *WEKA* library, requires to make use of the *WEKA* library in *PDM* as well.

4.3 Java Migration of Required Software and Libraries to Android

Migrating almost any open source *Java* application to the *Android* platform can be achieved with the procedure illustrated in Figure 4.3, which has been adopted for migrating *PDM* and its required libraries and software to *Android*.

This migration process starts by importing all the source code into an *Android* project. As *Android* also uses the *Java* language, most of the errors are caused by 'missing' functionality, other errors can be caused by *Android* specific requirements, like permissions. Fortunately 'Android includes a set of core libraries that provides most of the functionality available in the core libraries of the Java programming language'[8]. This means that you can expect to find the core *Java* functionality in *Android*. Thus the migration problem reduces to fixing these issues by following one or several of the of the following alternatives:

- If the functionality in question is not vital for the operation of the application, then it can be completely removed. For example, *JADE*'s framework offers a lot of functionality that might never be used by a particular application. It offers developers flexibility by providing a lot of functionality, but in the end they decide if they use it or not. Certainly defining which classes are dispensable is a hard task and should not be taken lightly. Some of the approaches to accomplish this

---

[8] http://developer.android.com/guide/basics/
what-is-android.html

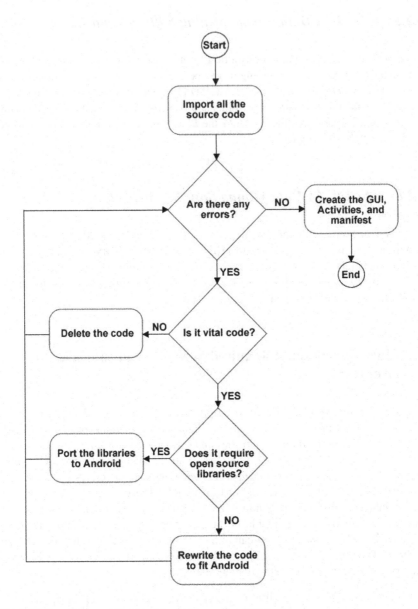

Fig. 4.3 Migration procedure used for implementing PDM

are: using development tools to browse for references in the project to the class in question, profiling the project to find references to that class or even experimenting by removing the class and looking at the impact it has on the application. This is not an exhaustive list, but shows some of the methods we used in the migration process.

- If the functionality is available as open source, it can be imported into the project. This might require more unimplemented functionality, thus introducing more problems, but at the end the application will resemble more its original implementation.

- If the functionality has a different approach in the *Android* platform, or if there are *Android* libraries that achieve the same action or comparable, then the application can be modified to adjust the functionality to the *Android* approach. For example, *Android* uses a completely different *GUI* approach, thus *GUI*s have to be rewritten to work under Android.

If there are no more remaining issues then *Android* specific chores like setting up the main activity, creating the necessary resources for the *GUI* and defining the manifest need to be performed.

Following this procedure we migrated all required software successfully to *PDM*, in particular we migrated *JADE* with reduced, but enough functionality to allow the implementation of mobile *PDM* agents for *Android* smartphones.

4.4   The PDM Implementation in the Mobile Environment

In this section, details about how *PDM* is implemented and deployed on-board smartphones running *Android OS* are provided.

4.4.1   Some Implementation Details

PDM's *GUI* has been developed for the user to communicate with *JADE*. *JADE*'s standard *GUI* does not work on *Android* devices, as *GUI*s in *Android* are conceptually handled differently compared with standard *Java*.

Basically *Android* relies on 'Activities' to represent a 'single, focused thing the user can do'[9]. Each activity can be seen as a 'frame' in *Java*, that is capable of displaying some content. As we are still developing *PDM* we have decided to show a simple console view to the user, where the user can look at all the output that *PDM* and *JADE* are generating, this view is updated every time there is new information to display.

In order to start up a *JADE* platform we have created an *Android* application that can start *JADE* on an *Android* device locally, the rest is then handled by *JADE*. The *JADE* platform connects to a different *JADE* instance on a different device and thus runs across different *Android* devices asynchronously. Also *JADE* is then in charge of doing whatever is necessary to keep the agents it hosts running and the connections between the *Android* devices alive.

---

[9] http://developer.android.com/reference/
android/app/Activity.html

JADE already offers several methods to create new agents. The easiest way is to use the *Remote Management Agent (RMA)* which presents a *GUI*. Unfortunately, to make this agent available under *Android*, the agent *GUI* would need to be rewritten to conform to *Android*'s user interface, and even redesigned to fit smaller screens. Certainly there are other ways to create agents, including the following:

- Sending a request directly to the *Agent Management System*;
- create the agent at start up, by sending it as an argument; or
- using a specialized agent to create other agents.

For the *PDM* implementation, a new agent by the name of *AgentCreator* was developed. The purpose of this agent is to control the creation of other agents, namely the *MADM* and *AM* agents discussed in Chapter 3. This is a very basic agent and only has two methods:

- **setup**: this method is called by *JADE* when the agent is initialized. In this case, it will store a reference to itself on the application. Thus the application is able to call the 'newAgent' method (listed below) of this agent, whenever it is needed.
- **newAgent**: this method takes as arguments the desired agent name, its class name and the passed parameters. Then it tries to create a new agent by first getting a reference to the container controller and the calling of the 'createNewAgent' method of *JADE*, with the variables obtained.

This *AgentCreator* agent is started automatically at the start of up of *JADE* on the local device. It must be noted that this would be done by every device on the network and that *JADE* does not support naming two agents identically. As these agents are located on different devices, the *IP* address could be used to name them differently. In the unlikely situation that the application is unable to determine the *IP* address, then a random number would be used. This alternative is not as good, as there is a small chance of name collision, while by definition, *IP* addresses can not be the same on different devices capable of communicating with each other.

4.4.2   Wireless Networks for PDM

Wireless communication using technologies such as *Wi-fi* and *Bluetooth* allows us to perform collaborative data mining techniques among these devices within the same range and running the same application. To get the devices connected, a network must be present. As *PDM* is intended for mobile phones, this network could be the carrier's network. The problem with this approach is that the use of *PDM* may incur higher costs derived from mobile data plans. The most common alternative is to use a wireless network. In the following, different scenarios for wireless network setup are given.

Scenario 1 - connecting to an existing network: a wireless network is already present and all the mobile phones that will participate in *PDM* have access to this network. This is the best possible scenario, as participants do not need to setup the network. Unfortunately, considering the moving nature of cell phones, wireless networks might become unreachable if the participants move away from the wireless hotspot.

Scenario 2 - bringing a wireless router: nowadays wireless routers are very compact in size and can be easily carried around. These routers only require electric power to operate, thus they only need a place with a spare working electric outlet. All the settings of the router (wireless band, password, and *IP* range) could be set beforehand and then it would just need to be connected to start acting as a hotspot.

Scenario 3 - using a phone as a wireless hotspot: one of the features present in recent smartphones is the ability to act as a wireless hotspot. This has great advantages when *PDM* is going to be executed 'on the move', but limits the amount of participants. Android currently supports up to 8 devices connected.

Scenario 4 - Wi-fi Direct (tm): The *Wi-fi* alliance has designed a new way to connect several devices using the existing wireless technology, named *Wi-fi Direct*[10]. This is already supported on the latest *Android* mobile phones. This scenario would work like the previous one, but makes the setup so much easier for the participants.

4.4.3   Establishing the PDM Network

Once all the participants have the application and the network is setup then the next step is to define a *PDM* server (mobile device or computer) and get all the participants connected to it. There are two options implemented to do this:

Option 1 - enter IP address: This could be used when the participants are located in different broadcast domains. For example, if the participants have opted to use Internet for *PDM*, then they would have to enter the *IP* address of the server.

Option 2 - broadcast messages: The intention of *PDM* is to have a powerful tool that can be used by anyone with little or no programming experience. In order to try to achieve this, one of the proposed ways of connecting is by broadcast messages. This goes with the assumption that local wireless networks would be used to establish the connection between the different participants. Thus, all of them would be in the same broadcast domain. The idea behind this is to have the server listening for broadcast messages on a particular port (we used port 15720). Whenever a new participant tries to join the *PDM* network, it would start sending broadcast messages until a server answers with its own *IP* address. The client then would automatically connect to the server and stop sending broadcast messages. This is assuming that there is only one server in the network. In future versions the server might send more information about its intentions, specifically on what it is working on and what it is trying to achieve with *PDM*. Then the client could gather a list of all the available servers on the network and present the user the different options available. Currently the application only starts sending these broadcast messages if

---

[10] http://www.wi-fi.org/

it determines that it is connected to a *Wi-fi* network. On the other hand, this approach would not work if it was connected through the carrier network.

4.5   Using PDM for Classification - A Walk through Example

This section demonstrates the usage of *PDM* using a classification example supported by screenshots of the actual *PDM* system. It is worth noting that the *PDM* prototype is still in development, hence some of the input parameters are currently hard coded but will be removed in future versions.

4.5.1   Starting PDM

The first action once *PDM* is started locally is to connect it to either an existing *PDM* network, or to create a new one. Connecting to an existing *PDM* network can be done automatically by querying the network and trying to find a device acting as a *PDM* server, or manually, by letting the user specify directly the *IP* address of the server. On the other hand, if there is no server at all, then the user could start a new *PDM* server. The *GUI* handling the start up of *PDM* is illustrated in Figure 4.4

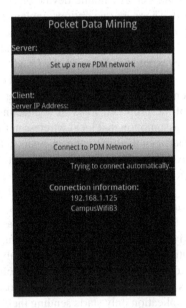

Fig. 4.4 The Android GUI for handling the start up of PDM or integrating a new device into an existing PDM network

Once the network has been setup and the devices are running on the same *PDM* platform, the user can start creating *AM* and *MADM* agents to perform data mining tasks in *PDM*.

4.5.2   Starting PDM Agents

Figure 4.5 shows the *Android GUI* for creating an *AM* agent in *PDM*. Note that the title of the *GUI* is 'Stream Miner Agent', which is our historic name for an *AM* agent. However, this change has not been incorporated in the current prototype yet.

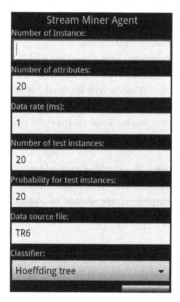

Fig. 4.5 The Android GUI for starting up a new AM on a mobile device

It should be noted that most of these parameters that can be inputted in the *GUI* are intended for test purposes, as they will allow *PDM* to simulate real world limitations. The input parameters are highlighted in the following:

- **Instance number:** This number is used to create a unique identifier ('AM' followed by the instance number) for each *AM* agent. Currently it must be assigned incrementally, starting with 1 and should not be repeated across the *PDM* network. However, *IP* addresses could be used to create a unique agent name similar to what has been done for the 'AgentCreator' agent highlighted in Section 4.4.1.
- **Number of attributes:** Currently all of the *AM* agents are sharing the same data source, but in a real scenario this might not be possible. In many cases, the agents will have access to different subsets of the information/data, as they have

different perspectives of the same data. In order to try to simulate this, instead of using all of the attributes of the data stream, only a random subset is chosen for each *AM*. The number of attributes that should be elected for each agent is determined by this parameter. It should be larger than 0 and not exceed the number of actual attributes in the dataset.

- **Data rate:** Not all the data instances used for *PDM* might be readily available, but instead they are being actively collected. For example, the continuous results of a sensor can not be determined from the beginning, but instead are being analyzed as soon as they become available. This behavior is simulated by using this parameter, which basically adds a constant time delay between readings. This parameter is expressed in milliseconds and should be greater than or equal to 0.
- **Number of test instances:** In order to try to calculate the accuracy of the classifier, some instances should be used for test purposes. This value determines the number of instances that will serve this task and that are chosen randomly from the available labeled instances in the dataset.
- **Probability for test instances:** This value should be between 1 and 100 and indicates the percentage with which instances should be picked as test instances. If it is 100, then all of the instances will be used for testing, until the number of test instances specified is met.
- **Data Source file:** Indicates the location of the data stream, which has originally been simulated from a text file for test purposes, hence the historic parameter name 'Data source file'.
- **Classifier:** This is the type of classifier embedded in the *AM*. Different types of *AM*s can produce varied results, so it is useful for experimenting with *PDM*. Some of the supported classifiers are shown in Figure 4.6. Recall that we used *Hoeffding Trees* and *Naive Bayes* for our experimental study reported in Chapter 3.

In *PDM* once *AM* agents are available, then *MADM* agents can be started to visit *AM* agents on different mobile devices within the same *PDM* network. We say that the *MADM* agents 'hop' from device to device. Figure 4.7 shows the *GUI* to start such an *MADM* agent. Note that the title of the *GUI* is 'Mobile Decision Agent', which is our historic name for an *MADM* agent. However, this change has not been incorporated in the current prototype yet.

Similar to the *GUI* for starting an *AM* agent, it should be noted that most of these parameters that can be inputted in the *GUI* are intended for test purposes, as they will allow *PDM* to simulate real world limitations. The input parameters are highlighted below:

- **Number of instance:** This value is used to generate the name that will uniquely identify the new *MADM*.
- **Data source file:** In the case of classification tasks this specifies the location of unlabeled data instances that need to be classified.
- **Weight threshold:** In order to make its decision, the *MADM* will use this value to determine if the weights (classification accuracy) returned by each *AM* are good enough to be considered.

Fig. 4.6 A list to the classifiers available to be used by an AM agent

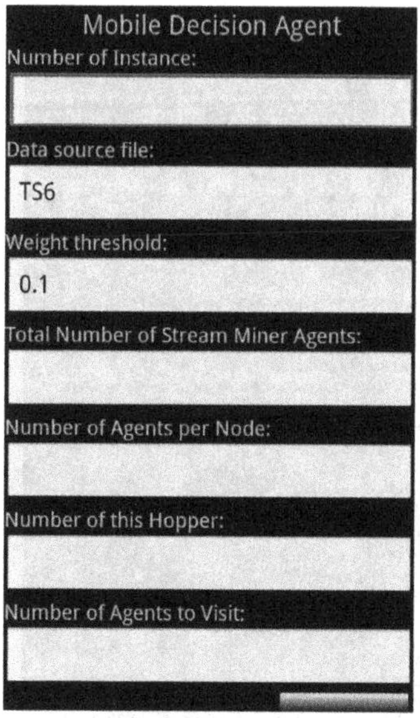

Fig. 4.7 The Android GUI for starting up a new MADM agent hopping through a PDM network

- **Total number of stream miner agents:** Indicates how many *AM* agents are known to be available on the network. The *MADM* will not necessarily try to visit all of them, but this value can be used to indicate the number of resources available, if known. This does not need to be specified but can be used to simulate real world limitations.
- **Number of agents per node:** Indicates the amount of *AM*s found on each mobile phone. This does not need to be specified but can be used to simulate real world limitations.
- **Number of this hopper:** This value is used for testing and debugging purposes and can be ignored here.
- **Number of agents to visit:** Explicitly states the amount *AM*s to be visited. This does not need to be specified but can be used to simulate real world limitations.

4.5.3   Retrieve Results from PDM Agents

In *Java* the console can be used to output information to the user. While *Android* supports this class, it does not offer a view of the console per se. Thus, the output never reaches the user. We changed this by redirecting the target of the output. Currently the main output screen of *PDM* resembles that of a console. An example output of PDM can be seen in Figure 4.8.

Fig. 4.8 The Android GUI for displaying PDM's output to the user

4.6   Limitations of the Current PDM Implementation

The *Mobile Agent Resource Discoverer (MRD)* has not been implemented yet, thus *PDM* requires that the actual number of *Agent Miners (AM)* is passed as a parameter to the *Mobile Agent Decision Maker (MADM)*. Consequently, all of the *AM*s should be initialised in the *JADE* network, following this naming convention: AM#, where # is an incremental value, starting from 1 that represents the number of the *AM*. In the case, an *AM* is missing from the network the *MADM* fails and is not able to recover. This behavior will be corrected when the *MRD* is implemented, which will be able to locate the different *AM*s disregarding their name. The current approach works just fine for evaluation purposes.

Another issue that should be addressed when the *MRD* is implemented is the fact that mobile devices could suddenly disconnect from the *JADE* network. The prevailing idea is that the *MRD* defines a schedule based on the requirements and the *AM*s found on the network, then the *MADM* just follows it. In the worst case scenario, the mobile device that currently holds the *MADM* could disconnect, removing the *MADM* from the *JADE* network. This is very unlikely to happen, nevertheless it is still possible. A mechanism should be devised to recover most of the progress obtained by the *MADM* and get it back on the network. However, the user has already the option to start several redundant and identical *MADM*s, which increases the chances of at least one *MADM* to return safely.

4.7   What Is Next?

Chapters 3 and 4 provided a comprehensive detailed information on *PDM* framework. In chapter 3, we provided the conceptual view of the framework along with experimental study proving the feasibility of the framework for classification tasks. In this chapter, we gave in sufficient details the recipe on how one can implement this framework in the mobile environment.

Although, up to this point, it seems that we provided all the necessary information about *PDM* framework, an important extension to *PDM* dealing with the problem of concept drift is provided in the following chapter. Concept drift is a data stream mining challenge that deals with the reality that models have to adapt to changes in the incoming data streams. Continuously built models provided by *PDM* do not take into consideration the possibility of occurring concept drifts. Owing to the importance of this problem in real-life application, we have extended *PDM* dealing with concept drifts.

Chapter 5
Context-Aware PDM (*Coll-Stream*)

5.1 Motivation

This chapter details how *PDM* can be extended to deal with the problem of *concept drift*. In such scenario, the goal is to learn an anytime classification model that represents the underlying concept from a stream of labeled records. Such a model is used to predict the label of the incoming unlabeled records. However, it is common for the underlying concept of interest to change over time and sometimes the labeled data available in the device is not sufficient to guarantee the quality of the results [66]. Therefore, we describe a framework that exploits the knowledge available in other devices to collaboratively improve the accuracy of local predictions.

The data mining problem is assumed to be the same in all the devices, however, the feature space and the data distributions may not be static as assumed in traditional data mining approaches [60, 98]. We are interested in understanding how the knowledge available in the community can be integrated to improve local predictive accuracy in a ubiquitous data stream mining scenario, which is the main objective of *PDM*.

As an illustrative example, collaborative spam filtering [27] is one of the possible applications for the collaborative learning framework. Each ubiquitous device learns and maintains a local filter that is incrementally learned from a local data stream based on features extracted from the incoming mails. In addition, user usage patterns and feedback are used to supervise the filter that represents the target concept (i.e., the distinction between spam and ham). *Mobile Activity Recognition* [46, 47] can be another interesting application. The ubiquitous devices collaborate by sharing their knowledge with others, which can improve their local device predictive accuracy. Furthermore, the dissemination of knowledge is faster, as peers new to the mining task, or that have access to fewer labeled records, can anticipate spam patterns that were observed in the community, but not yet locally. Moreover, the privacy and computational issues that would result from sharing the original mail are minimized, as only the filters are shared. Consequently, this increases the efficiency of the collaborative learning process.

M.M. Gaber, F. Stahl, and J.B. Gomes, *Pocket Data Mining*, Studies in Big Data 2,   61
DOI: 10.1007/978-3-319-02711-1_5, © Springer International Publishing Switzerland 2014

However, many challenges arise from this scenario, in the context of *PDM* the two major ones are:

1. how the knowledge from the community can be exploited to improve local predictiveness; and
2. how to adapt to changes in the underlying concept.

To address these challenges, we will describe *Coll-Stream*, an incremental ensemble approach where the models from the community are selected and weighted based on their local accuracy for different partitions of the instance space. This technique is motivated by the possible conflicts among the community models. The technique allows to exploit the fact that each model can be accurate only for certain subspaces, where its expertise matches the local underlying concept.

We should note that the communication costs and protocols to share models between devices are out of the scope of this chapter and represent an interesting open challenge.

5.2   Problem Definition

Let X be the space of attributes and its possible values and Y be the set of possible (discrete) class labels. Each ubiquitous device aims to learn the underlying concept from a stream DS of labeled records where the set of class labels Y is fixed. However, the feature space X does not need to be static. Let $X_i = (\mathbf{x_i}, y_i)$ with $x_i \in X$ and $y_i \in Y$, be the ith record in DS. We assume that the underlying concept is a function f that assigns each record x_i to the true class label y_i. This function f can be approximated using a data stream mining algorithm to train a model m on a device from the DS labeled records. The model m returns the class label of an unlabeled record $\mathbf{x}$, such that $m(\mathbf{x}) = y \in Y$. The aim is to minimize the error of m (i.e., the number of predictions different from f). However, the underlying concept of interest f may change over time and the number of labeled records available for that concept can sometimes be limited. To address such situations, we propose to exploit similarities in models from other devices and use the available labeled records from DS to obtain the model m. We expect m to be more accurate than using the local labeled records alone when building the model. The incremental learning of m should adapt to changes in the underlying concept and easily integrate new models. We assume that the models from other devices are available and can be integrated at anytime. The costs and methods used to generate and share models are beyond the scope of this work.

5.3 Collaborative Data Stream Mining

In collaborative and distributed data mining, the data is partitioned and the goal is to apply data mining to different, usually very small and overlapping, subsets of the entire data [28, 93]. In this work, our goal is not to learn a global concept, but to learn from other devices their concepts, while maintaining a local or subjective point of view. Wurst and Morik [105] explore this idea by investigating how communication among peers can enhance the individual local models without aiming at a common global model. The motivation is similar to what is proposed in domain adaptation [29] or transfer learning [73]. However, these assume a batch scenario. When the mining task is performed in a ubiquitous environment [36, 55], an incremental learning approach is required.

In ubiquitous data stream mining, the feature space of the records that occur in the data stream may change over time [60] or be different among devices [105]. For example, in a stream of documents where each word is a feature, it is impossible to know in advance which words will appear over time, and thus what the best feature space to represent the documents with is. Using a very large vocabulary of words is inefficient, as most of the words will likely be redundant and only a small subset of words is finally useful for classification. Over time it is also likely that new important features appear and that previously selected features become less important, which brings change to the subset of relevant features. Such change in the feature space is related to the problem of *concept drift*, as the target concept may change due to changes in the predictiveness of the available features.

The following sections describe the details of *Coll-Stream* and Figure 5.1 illustrates the collaborative learning process on *PDM*.

5.4 Augmenting PDM with Dynamic Classifier Selection

Coll-Stream is a collaborative learning approach for ubiquitous data stream mining that combines the knowledge of different models from the community.

There is a large number of ensemble methods to combine models, which can be roughly divided into:

1. voting methods, where the class that gets more votes is chosen [96, 100, 61]
2. selection methods, where the 'best' model for a particular instance is used to predict the class label [107, 99].

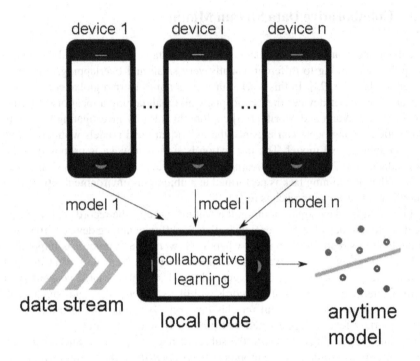

Fig. 5.1 Collaborative learning process

Data: Data stream DS of labled records, window w of records.
repeat
  Add next record DS_i from DS to w;
  if $w \to numRecords > wMaxSize$ **then**
   $forget(w \to oldestRecord)$;
  end
  $r = getRegion(DS_i)$;
  ForAll$Model \to m_j$;
  $prediction := m_j.classify(DS_i)$;
  if $prediction = DS_i \to class$ **then**
   $updateRegionCorrect(r, m_j)$;
  end
  $updateRegionTotal(r, m_j)$;
until *END OF STREAM*;

Algorithm 2. *Coll-Stream* Training

Data: Data stream DS of unlabeled records.
repeat
    Get DS_i from DS;
    $r := getRegion(DS_i)$;
    forall the $Model \to m_j$ **do**
        $model := argmax_j(getAccuracy(m_j, r))$;
    end
    return $model.classify(DS_i)$;
until *END OF STREAM*;

Algorithm 3. *Coll-Stream* Classification

The *Coll-Stream* is a selection method that partitions the instance space X into a set of regions R. For each region, an estimate of the models accuracy is maintained over a sliding window. This estimated value is updated incrementally as new labeled records are observed in the data stream or new models are available. This process (in Algorithm 2), can be considered a *Meta-Learning* task where we try to learn for each model from the community how it best represents the local underlying concept for a particular region $r_i \in R$. When *Coll-Stream* is asked to label a new record $\mathbf{x_i}$, the best model is used. The best model is considered to be the one that is more accurate for the partition r_i that contains the new record, as detailed in Algorithm 3. The accuracy for a region r_i is the average accuracy for each partition of its attributes. For r_{15} in Figure 5.2, we average the accuracy for value $V1$ of attribute $A1$ and value $V5$ of attribute $A2$. The accuracy is the number of correct predictions divided by the total number of records observed (these values are updated in lines 10 and 12 of Algorithm 2) as illustrated in Figure 5.2. The next section explains how the regions are created using the attribute values.

5.4.1 Creating Regions

An important part of *Coll-Stream* is to learn for each region of the instance space X which model m_j performs better. This way m_j predictions can be used with confidence to classify incoming unlabeled records that belong to that particular region.

The instance space can be partitioned in several ways. Here we follow the method used by [107], where the partitions are created using the different values of each attribute. For example, if an attribute has two values, two estimators of how the classifiers perform for each value are kept. If the attribute is numeric, it is discretized and the regions use the values that result from the discretization process. This method has shown good results and it represents a natural way of partitioning the instance space. However, there is an increased memory cost associated with a larger number of regions. To minimize this cost, these regions can be partitioned into higher granularity ones, aggregating attribute values into a larger partition. This is illustrated in Figure 5.2, where the values $V4$ and $V5$ of attribute $A1$ are grouped into regions $r41$ to $r45$.

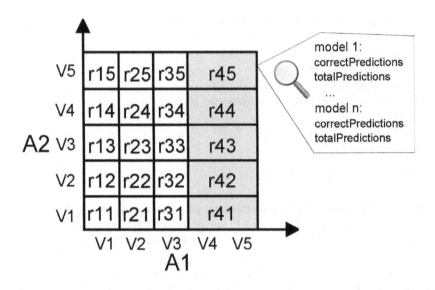

Fig. 5.2 Detail of the estimators maintained for each region

Figure 5.3 illustrates the training and classification procedures of *Coll-Stream* that are described in Algorithm 2 and Algorithm 3.

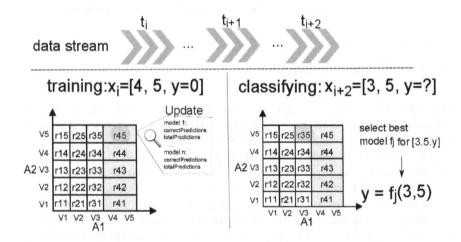

Fig. 5.3 Coll-Stream: Training and Classifying

5.4.2   *Variations*

Some variations of the *Coll-Stream* approach can be considered:

Multiple classifier selection:

If more than one model is selected, their predictions are weighed according to the corresponding accuracy for region r_i, and the record to be labeled gets the class with the highest weighted vote. This is similar to weighted majority voting but with a variable number of classifiers, where the weights are calculated in relation to the region that contains the unlabeled record to classify.

Feature weighting:

The models used from the community can represent a heterogeneous feature space as each one is trained according to a different data stream DS_d. One possible variation is for each device to measure feature relevance. Then at the time of classification the accuracy estimates for each region are weighted according to the feature weight for that region. The predictive score of each feature can be computed using popular methods such as, the information gain, χ^2, or mutual information [60]. However, these must be calculated incrementally given the data stream scenario where this approach is framed. Moreover, this takes into account that features that were relevant in the past can become irrelevant at some point in the future for a different target concept.

Using local base learner:

One base learner that is trained using the available records in the device can be always part of the ensemble. This integration is simple as it only requires an additional step of training the classifier when a new record arrives in addition to updating the ensemble estimates for the new record region.

Resource awareness:

Resource-awareness is an important issue in ubiquitous data stream mining [36, 40]. In such a dynamic ubiquitous usage scenario, it is common for the *Coll-Stream* method to receive too much knowledge from the community over time. In such situations we propose to discard past models that have the lowest performance and allow the integration of new models.

Feature selection:

Feature selection is used to reduce the size of the models kept. This process can be executed before the models are shared, in order to additionally reduce the communication costs associated with transferring the models between devices. When the model is shared with other devices, only its most predictive features are used. For example in the case where the *Naive Bayes* algorithm is used as base learner, only the corresponding estimators for the selected predictive attributes given the class take part in the shared model. The feature's predictiveness is evaluated and these are then selected. However, many selection methods can be used for this task and according to the application some methods may be superior to others. Some simple possibilities are:

- Fixed N, select the top N highly scored features for each model.
- Fixed Threshold, which defines the cut point between predictive and irrelevant features.

Context-awareness:

Context awareness is an important part of ubiquitous computing [70]. In most ubiquitous applications concepts are associated with context, this means that they may reappear when a similar context occurs. For example, a weather prediction model usually changes according to the seasons. The same applies with product recommendations or text alerting models where the user interest and behavior might change over time due to fashion, economy, spatial-temporal situation or any other context [102]. This has motivated some works [48, 13] to analyse how to use context to track concept changes.

Particularly in context-aware *Coll-Stream* [12], context information is associated with the models kept. At the time of classifying unlabeled records, the model that is more accurate for the partition of the feature space that contains the unlabeled record and that is associated with a context similar to the current one.

5.5   Discussion

This chapter discusses collaborative data stream mining in ubiquitous environments and describes *Coll-Stream*, an ensemble approach that incrementally learns which classifiers from an ensemble are more accurate for certain regions of classification problem the feature space. *Coll-Stream* is able to adapt to changes in the underlying concept using a sliding window of the classifier estimates for each region. Moreover, we also discussed the possible variations of *Coll-Stream*.

Coll-Stream represents an important extension to the basic *PDM* framework to deal with the concept drift problem. Proving the feasibility of the approach, the following chapter provides a thorough experimental study.

Chapter 6
Experimental Validation of Context-Aware PDM

Having discussed in sufficient details our *Coll-Stream* technique extending *PDM* to deal with the concept drift problem, we conducted experiments to test the proposed approach's accuracy in different situations, using a variety of synthetic and real datasets. The implementation of the learning algorithm was developed in *Java*, using the *MOA* [19] environment as a test-bed. The *MOA* evaluation features and some of its algorithms were used, both as base classifiers to be integrated in the ensemble of classifiers and in the experiments for accuracy comparison.

6.1 Datasets

A description of the datasets used in our experimental studies is given in the following.

6.1.1 STAGGER

This dataset was introduced by Schlimmer and Granger [83] to test the *STAGGER* concept drift tracking algorithm. The *STAGGER* concepts are available as a data stream generator in *MOA* and has been used as a *benchmark* dataset to test concept drift [83]. The dataset represents a simple block world defined by three nominal attributes *size*, *color* and *shape*, each with 3 different values. The target concepts are:

- $size \equiv small \wedge color \equiv red$
- $color \equiv green \vee shape \equiv circular$
- $size \equiv (medium \vee large)$.

M.M. Gaber, F. Stahl, and J.B. Gomes, *Pocket Data Mining*, Studies in Big Data 2,   69
DOI: 10.1007/978-3-319-02711-1_6, © Springer International Publishing Switzerland 2014

6.1.2 SEA

The *SEA* concepts dataset was introduced by Street and Kim [96] to test their *Stream Ensemble Algorithm*. It is another *benchmark* dataset as it uses different concepts to simulate concept drift, allowing control over the target concepts in our experiments. The dataset has two classes {class 0, class 1} and three features with values between 0 and 10 but only the first two features are relevant. The target concept function classifies a record as class 1 if $f_1 + f_2 \leq \theta$ and otherwise as class 0. The features f_1 and f_2 are the two relevant ones and θ is the threshold value between the two classes. Four target concept functions were proposed in [96], using threshold values 8, 9, 7 and 9.5. This dataset is also available in *MOA* as a data stream generator, and it allows control over the noise in the data stream. The noise is introduced as the $p\%$ of records where the class label is changed.

6.1.3 Web

The *webKD* dataset[1] contains web pages of computer science departments of various universities. The corpus contains 4,199 pages (2,803 training pages and 1,396 testing pages), which are categorized into: *project*; *course*; *faculty*; *student*. For our experiments, we created a data stream generator with this dataset and defined 4 concepts, that represent user interest in certain pages. These are:

- *course* ∨ *project*
- *faculty* ∨ *project*
- *course* ∨ *student*
- *faculty* ∨ *student*

6.1.4 Reuters

The *Reuters* dataset[1] is usually used to test text categorization approaches. It contains 21,578 news documents from the *Reuters* news agency collected from its newswire in 1987. From the original dataset, two different datasets are usually used, R52 and R8. R52 is the dataset with the 52 most frequent categories, whereas R8 only uses the 8 most frequent categories. The R8 dataset has 5,485 training documents and 2,189 testing documents. In our experiments from R8, we use the most frequent categories: *earn* (2,229 documents), *acq* (3,923 documents) and *others* (a group with the 6 remaining categories, with 1,459 documents). Similar to the *Web* dataset, in our experiments, we define 4 concepts (i.e., user interest) with these categories. These are:

---

[1] http://www.cs.umb.edu/ smimarog/textmining/
datasets/index.html

- *others*
- *earn*
- *acq*
- *earn* ∨ *others*

6.2 Experimental Setup

We test the proposed approach using the previously described datasets with the data stream generator in *MOA*, the target concept was changed sequentially every 1,000 records and the learning period shown in the experiments is of 5,000 records. This number of records allows for each of the concepts to be seen at least once for all the datasets used. In addition, for each concept in all the datasets using 1,000 records is more than required to observe a stable learning curve. As parameters, we used for the window size 100 records and this was fixed for all the experiments and for all the algorithms that use a *sliding window*. This guarantees the robustness of the approach without fine parameter tuning, which is a drawback of other existing approaches. The influence of such parameter on the results is contrasted with a version of the *Naive Bayes* algorithm over a sliding window. Consequently, we can distinguish the gains coming from the collaborative ensemble approach and the ones coming from the adaptation of using a sliding window.

The approaches compared in the experiments are:

- *Coll-Stream*, the approach proposed in this work.
- *MajVote*, Weighted Majority Voting, ensemble approach where each classifier accuracy is incrementally estimated based on its predictions. To classify a record, each classifier votes with a weight proportional to its accuracy. The class with most votes is used.
- *NBayes*, incremental version the *Naive Bayes* algorithm.
- *Window*, incremental Naive Bayes algorithm but its estimators represent information over a sliding window;
- *AdaHoeffNB*, Hoeffding Tree that uses an adaptive window (*ADWIN*) to monitor the tree branches and replaces them with new branches when their accuracy decreases [18].

In addition, we have implemented and tested the accuracy of *Coll-Stream* variations proposed in Section 5.4.2. Nevertheless, the results show only a very small increase in accuracy for the variation that considers the relative importance of the features and are not significant for the other variations. For this reason the results presented in this section refer to the standard version of *Coll-Stream* and Section 6.7 is dedicated to describe the experiments of the variation that uses feature selection.

In the experiments, the base classifiers (that represent the community knowledge) used in the ensemble were trained using data records that correspond to each individual concept, until a stable accuracy was reached. We used the *Naive Bayes* and *Hoeffding Trees* algorithms available in *MOA* as base classifiers, for consistency with experiments reported in Chapter 3 for the base version of *PDM*. Therefore, for each concept the ensemble receives 2 classifiers. For the real datasets, the ensemble only receives the 3 first of the 4 possible concepts, this asserts how the approach is able to adapt the existing knowledge to a new concept that is not similar to the ones available in the community (note that for each still two base classifiers are received). In the experiments we record the average classification accuracy over time using a time window of 50 records using the evaluation features available in *MOA*. In the synthetic datasets we tested different seeds to introduce variability in the results, but because of the large number of records the classifiers easily capture the target concept without the seed causing any significant influence on the accuracy.

6.3   Accuracy Evaluation of *Coll-Stream*

We compare the efficacy of *Coll-Stream* in relation to the other aforementioned approaches. In this set of experiments we measured the predictive accuracy over time. The vertical lines in the figures indicate a change of concept.

Table 6.1 Accuracy evaluation

DataSet	STAGGER	SEA	Web	Reuters
AdaHoeffNB	78.86%	89.72%	58.24%	68.08%
NBayes	72.74%	90.96%	57.06%	62.90%
Window	81.96%	92.42%	58.62%	72.36%
MajVote	93.76%	90.98%	66.16%	66.94%
Coll-Stream	**97.42%**	**94.72%**	**71.00%**	**76.92%**

Table 6.1 shows the overall accuracy of the different approaches for each dataset. Concerning the accuracy, *Coll-Stream* consistently achieves the highest accuracy. For the other approaches, the performance seems to vary across the datasets. The real datasets are very challenging for most of the approaches.

Figure 6.1 depicts further analysis of the accuracy of the different approaches over time for the *STAGGER* data. It shows that *Coll-Stream* is not only the more accurate but also the most stable approach, even after concept changes. The *MajVote* also achieves very good results, close to *Coll-Stream*, but for the 2nd and 3rd concept it performs worse than *Coll-Stream*. For the *Window*, *AdaHoeffNB* and *NBayes* approaches, the first is able to adapt faster to concept drift, while *AdaHoeffNB* only shows some gain over the *NBayes*, which is the worst approach in this evaluation, due to the lack of adaptation.

Figure 6.2 shows the high and stable accuracy of *Coll-Stream* over time for the *SEA* data. In this experiment, we can observe that the *MajVote* performs worse than *Coll-Stream*, and do similarly the other methods with the exception of the 4th concept, where the *MajVote* achieves the best performance. The *Window* approach

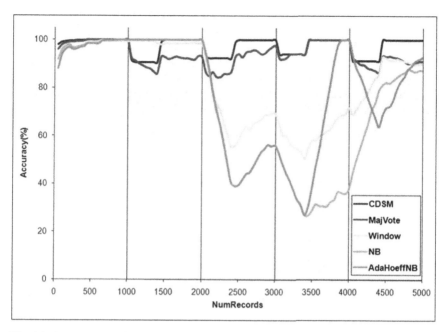

Fig. 6.1 Accuracy over time for the STAGGER datastream

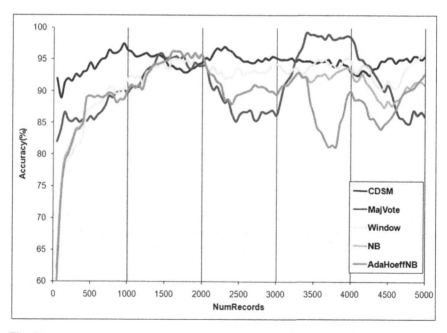

Fig. 6.2 Accuracy over time for the SEA datastream

also shows good accuracy and stable performance with the changing concept, which makes it higher than *MajVote* when we look at the overall accuracy in Table 6.1. The *NBayes* and *AdaHoeffNB* approaches do not show significant difference. We should note that even the *AdaHoeffNB*, which achieves the worse performance in the evaluation, is able to keep the accuracy higher than 80%. This can be a result of less abrupt differences between the underlying concepts, when compared with what we observed in the *STAGGER* data in Figure 6.1.

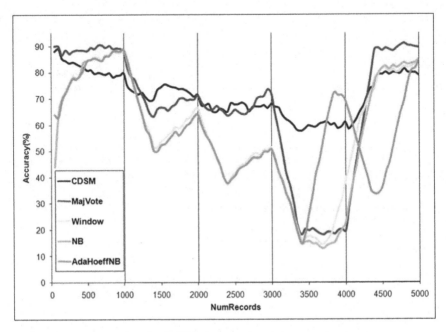

Fig. 6.3 Accuracy over time for the Web data stream

The *Web* data concepts are more complex than the ones that exist in the synthetic data. For this reason, we can observe in Table 6.1 that the overall accuracy is not as high for most of the approaches. Figure 6.3 further analyses the accuracy curve for the different concepts and how it is affected by concept changes. For the 1st concept, the *MajVote* achieves a slightly better performance than *Coll-Stream*. However, in the 2nd concept we can observe a greater drop in the performance of *MajVote* at the time that the *Coll-Stream* is more stable and become higher in the accuracy. During the 3rd concept, both approaches achieve similar results, while the other approaches are not able to adapt successfully to the concept changes. It is interesting to find that for the 4th concept, which is dissimilar to the classifiers used in the ensemble approaches, we observe that *Coll-Stream* is able to adapt well with only a slight drop in accuracy while the *MajVote* shows a large drop in performance and is not able to adapt successfully. Again when the 1st concept recurs we see a dominance of the *MajVote* which seems to represent this concept with high accuracy.

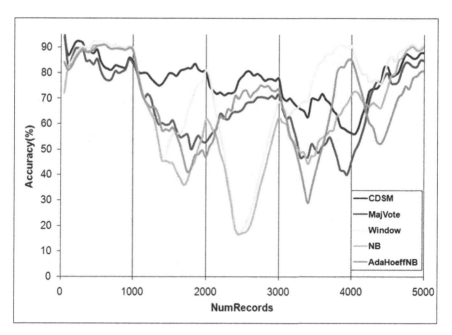

Fig. 6.4 Accuracy over time for the Reuters datastream

Using the *Reuters* dataset, the overall accuracy is better than in the *Web* data as can be observed in Table 6.1. Figure 6.4 shows that *Coll-Stream* achieves high accuracy across the concepts and very stable performance over time. The results are somehow similar to the *Web* ones. However, the *MajVote* achieves worse performance for the 2nd and 3rd concepts, while the *Window* approach is able to adapt faster to the different concepts, with the exception of the 3rd one. For the 4th concept, the *Window* approach is even able to outperform *Coll-Stream*, which explains its overall accuracy in Table 6.1. The *AdaHoeffNB* also is able to adapt to the concept drift but this adaptation is not as fast as *Coll-Stream*.

6.4   Impact of Region Granularity on the Accuracy

In this set of experiments we measured how the accuracy is influenced by the granularity of the partitions used in *Coll-Stream*. For the *SEA* dataset where each attribute can take values between 0 and 10, we defined the regions with different sizes from 10 possible values to only 2 values for each attribute. For example, if we consider *R2*, the 2 indicates that each attribute has to be discretized into only two values. Consequently, all the accuracy estimations for attribute values greater than or equal to 5 are stored in one region, while values lower are stored in another. Figure 6.5 shows that the accuracy of *Coll-Stream* decreases with higher region granularity

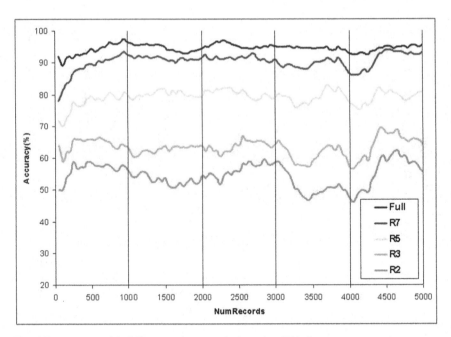

Fig. 6.5 Accuracy with different region granularity using SEA data stream

(i.e., less partitions). In Table 6.2 we measured the memory required for the different granularities and how that size relates to the overall memory consumption of the approach (excluding the classifiers). We observe that the additional memory cost to have higher accuracy is small. This could only have a significant impact in ubiquitous devices with very limited memory, where the accuracy-efficiency trade-off of the approach is critical. This is an important requirement for *PDM*.

Table 6.2 Region granularity evaluation using SEA dataset

Regions	Accuracy	Memory(bytes)	Memory(%)
Full	94.72%	30112	55.83%
R7	90.82%	23776	44.09%
R5	79.58%	20608	38.21%
R3	63.34%	17440	32.33%
R2	54.64%	15856	29.40%

The results show that *Coll-Stream* can work in situations with memory constraints and still achieve a good trade-off between accuracy and the memory consumed. It can be seen that R5 and R7 are competitive while at the same time saving around 50% memory consumption. The resource efficiency of such approaches to ubiquitous knowledge discovery also opens additional issues for future research work. Particularly, when exploring other representation strategies that can save memory and will also result in lower communication overhead.

6.5   Impact of Noise on the Accuracy

We compare the impact of noise on the accuracy of *Coll-Stream*. Table 6.3, shows the results of our experiments with different approaches using the *SEA* data with different noise percentages (i.e., percentage of records where the class label changed). The first column represents the case without noise and shows the results that were previously reported in Section 6.3. We can observe that *Coll-Stream* achieves higher accuracy than the other approaches even when the noise level increases. However, as the percentage of noise increases the difference between the approaches decreases. Consequently, when the noise level is 30%, all of the approaches achieve a very similar performance (around 63%).

Table 6.3 Noise impact evaluation using SEA datastream

DataSet	Noise 0%	Noise 10%	Noise 20%	Noise 30%
AdaHoeffNB	89.72%	81.08%	71.44%	63.12
NBayes	90.96%	81.94%	72.72%	**64.12**
Window	92.42%	82.82%	73.12%	63.90
MajVote	90.98%	81.42%	71.96%	63.90
Coll-Stream	**94.72%**	**83.68%**	**73.22%**	63.78

6.6   Effect of Concept Similarity in the Ensemble

In Section 6.3, when discussing the evaluation of the experiments using real datasets, we were able to observe (in Figures 6.3 and 6.4) that *Coll-Stream* is able to adapt to new concepts that are not represented in the community/ensemble. This is clear when we compare *Coll-Stream's* performance difference with the *MajVote* for the 4th concept (between 4,000 and 5,000 records) in the real datasets. To further investigate this issue, we performed an additional experiment where we measured the impact on the accuracy of *Coll-Stream* when having the target concept represented in the ensemble. We can observe the results in Table 6.4. The table shows a small drop

Table 6.4 Similarity with Target Concept (TC)

DataSet	Without TC	With TC
STAGGER	95.86%	97.42%
SEA	94.24%	94.72%
Web	71.00%	72.36%
Reuters	76.92%	77.78%

in accuracy between the two cases; when the target concept is represented and when it is not. Thus, it could be concluded that *Coll-Stream* achieves good adaptation to new concepts using existing ones. Furthermore, for the *SEA* dataset we observe the least difference, because even without knowledge from the 4th concept, there is greater similarity to known ones than in other datasets (e.g., in the *STAGGER* dataset where the difference between concepts across the regions is greater). Consequently, if there is a local similarity among the concepts, *Coll-Stream* is able to exploit it. This way it can represent a concept by combining other concepts that are locally similar to the target one.

6.7   Impact of Feature Selection on the Accuracy

In general, accuracy evaluation of *Coll-Stream* when using feature selection shows that it is possible to maintain or even increase the accuracy, while reducing the number of features that need to be kept. This has a strong impact on the accuracy-efficiency trade-off of the approach and will be discussed in detail in the following subsection where we evaluate the memory consumption of *Coll-Stream*. In addition, we observe in Table 6.5 that there is a small decrease in the accuracy of *Coll-Stream*, particularly for the *Web* dataset, in this set of experiments in relation to the experiments in the previous section. This is a result of using less diversity in the ensemble (i.e., only models from *Naive Bayes* as base learner are used).

Table 6.5 shows the 5 different tested methods, their parameters for each dataset and the accuracy obtained. In what concerns the accuracy for the different datasets, we can observe in Table 6.5 that for the synthetic datasets, where the number of features is much smaller than in the real datasets. Therefore, it is only possible to perform a modest reduction on the number of features without affecting the accuracy. This is also a consequence of the number of irrelevant features. For instance, in the *STAGGER* dataset the number of irrelevant features can be 1 or 2 according to the target concept. Moreover, in the *SEA* dataset the last feature is always irrelevant to the target concept, we can observe that when the number of kept features is two (and the feature selection method correctly selects the two predictive ones) the accuracy increases. Nevertheless, if one of the predictive features is lost there is a sharp drop in accuracy.

For the real datasets, where there is a large number of features the results show that it is possible to reduce the number of features while achieving a similar or slightly better accuracy than without feature selection.

With respect to the different feature selection methods, either the fixed or threshold approaches achieve similar results. However, the main drawback associated with this method is related to the selection of the appropriate parameter value (i.e., threshold or number of features). In general, the fixed approach allows better control over the consumed space, while the threshold approach is more flexible. This will be further analyzed in the next section where we asses the memory savings that result from each method.

Table 6.5 Accuracy evaluation of *Coll-Stream* using feature selection

DataSet	Measure	Fixed(1)	Fixed(2)	Threshold(1)	Threshold(2)	WithoutFS
STAGGER	Accuracy	90.78%	**97.40%**	93.98%	**97.40%**	**97.40%**
	ParValue	1	2	0.13	0.05	-
SEA	Accuracy	87.78%	**96.12%**	95.10%	**96.12%**	94.72%
	ParValue	1	2	0.08	0.06	-
Web	Accuracy	67.38%	66.54%	**67.80%**	66.54%	66.4%
	ParValue	100	300	0.08	0.02	-
Reuters	Accuracy	**77.00%**	76.32%	76.68%	76.06%	75.64%
	ParValue	100	300	0.08	0.02	-

6.8   Impact of Feature Selection on Memory Consumption

When measuring the savings in memory consumption that result from using feature selection, we can observe in Table 6.6 that it is possible to maintain or even increase the accuracy, while consuming at least 50% or less of the resources. Please observe this from the number of features (*NumF*) and the percentage of memory (*Mem*) used in relation to the test without feature selection (*WihoutFS*).

Table 6.6 Memory evaluation of *Coll-Stream* using feature selection

DataSet	Measure	Fixed(1)	Fixed(2)	Threshold(1)	Threshold(2)	WihoutFS
STAGGER	Accuracy	90.78%	**97.40%**	93.98%	**97.40%**	**97.40%**
	NumF	3	6	4	**5**	9
	Memory	33%	66%	44%	**55 %**	100 %
SEA	Accuracy	87.78%	**96.12%**	95.10%	**96.12%**	94.72%
	NumF	4	8	5	8	12
	Memory	33%	66%	42%	66%	100 %
Web	Accuracy	67.38%	66.54%	**67.80%**	66.54%	66.4%
	NumF	300	900	**234**	782	2820
	Memory	14%	43%	**11%**	37%	100 %
Reuters	Accuracy	**77.00%**	76.32%	76.68%	76.06%	75.64%
	NumF	300	900	502	986	1683
	Memory	18%	54%	30	59 %	100 %

6.9   Discussion

In this chapter, we present the results of the evaluation of *Coll-Stream*. Several experiments were performed using 2 known datasets for concept drift and 2 popular datasets from text mining from which we created a stream generator. We tested and compared *Coll-Stream* with other related methods in terms of accuracy, noise, partition granularity and concept similarity in relation to the local underlying concept.

The experimental results show that the *Coll-Stream* approach mostly outperforms the other methods and could be used for situations of collaborative data stream mining as it is able to exploit local knowledge from other concepts that are similar to the new underlying concept.

This chapter brings a conclusion to the techniques developed to realize the *PDM* framework. Potential applications of this framework in a variety of domains are discussed in the following chapter.

Chapter 7
Potential Applications of Pocket Data Mining

Potential applications for *PDM* are manifold and comprise amongst others the areas of financial investments, the health sector, public safety and defence. This chapter introduces possible application scenarios in these areas, however, many more are possible, especially in science, such as human behavior detection, user behavior detection in order to detect the use of the phone by a potential thief, or detection of potential for mass panics on large public events.

7.1 PDM as A Decision Support System for Stock Market Brokers

7.1.1 MobiMine

Recall that in *MobiMine* discussed in Chapters 2 and 3, a stock market broker uses his/her *Personal Digital Assistant (PDA)* or smartphone to subscribe to stock market data streams s/he is interested in [76]. *MobiMine* composes a watch list of interesting shares out of the data. For example, if the user is interested in computer technology shares, s/he would subscribe to a subset of the stream that comprises (only) these shares. *MobiMine* constantly updates the watch list of shares that are or may become interesting in the near future. For example, if in the past share A has made a huge profit, then share B usually made a huge profit as well. Consequently, in the future *MobiMine* may predict that if share A made huge profits then share B is likely to make huge profits as well, however, share B's profit may be slightly delayed to the profit increase of A. If this is the case *MobiMine* will put share B into the broker's watch list, and thus draws the brokers attention to share B. The broker can than use its expert knowledge and experience to decide to invest or not to invest into this share, however, there are many more reasons a share might be interesting.

M.M. Gaber, F. Stahl, and J.B. Gomes, *Pocket Data Mining*, Studies in Big Data 2, DOI: 10.1007/978-3-319-02711-1_7, © Springer International Publishing Switzerland 2014

7.1.2   How PDM Could Address Shortcominings of MobiMine

The shortcoming of *MobiMine* is that it only explores data that the user subscribed to (for example, only computer technology shares). However, it does not address that the currently subscribed shares may be influenced by non subscribed shares. For example, processors are made of gold, the gold price will have a direct impact on computer technology companies that produce computer chips, which in turn may have influence on computer technology companies that produce software. However, if the broker only subscribes to computer technology shares, the influenced companies will not be put into the watch list, which is a limitation of *MobiMine*. This is just one example showing how non subscribed companies may have impact on subscribed ones. One can argue, 'why does the user not simply subscribe to all of the shares'. In the following, we reiterate the reasons we briefly discussed in Chapter 3 of this book.

- Bandwidth consumption is a critical factor in mobile phone connections and is further deteriorating as more and more smartphones are sold and in use. Also larger bandwidth consumption will drain the battery far more quickly.
- Power consumption is critical as local algorithms have to do more work as there is more data to process.
- There may be subscription fees for certain parts of the data.

7.1.2.1   PDM Addressing MobiMine Shortcomings

PDM could build a local watch list using the local share subscription and subscriptions of different brokers. The local watch list is created by a local *Agent Miner (AM)* that could implement the algorithms of *MobiMine* or different more customized ones. These *AM*s create a local watch list for the broker the *AM* belongs to. However, as shown in Figure 7.1 global influence could be incorporated by sending a *Mobile Agent Decision Maker (MADM)* that consults local *AM*s in order to find global relationships to the subscribed data. Also users may have subscribed to overlapping shares. Lets say broker X is interested in shares A, B, C and user Y is interested in shares B,D,E. Now X sends an *MADM* to all participating mobile devices among which is Y's mobile device. The *MADM* discovers that Y's *AM* has put B and E into the watch list, thus the *MADM* may advise X to put B into its watch list as X is interested in share B.

7.1.2.2   PDM as a Decision Support System for Investments

If a broker is interested in a new share, but has no experience in investing in this particular share, or the market in which the share is traded, then s/he may use *PDM* in order to support its investment decision. For example, in a *PDM* network of brokers, a broker could run a classification system wrapped in an *AM*, if the broker

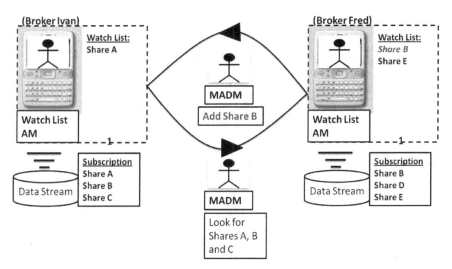

Fig. 7.1 Updating the watch list with information from brokers that have subscribed to different parts of the stock market data

makes a decision to buy or sell, the classifier model is updated accordingly. Shown in Figure 7.2, the broker who is unsure about investing in a new share may use an *MADM* to visit other brokers' *AMs*, collect the predicted investment decisions of other possibly more experienced brokers, and use them to support its own decision for the investment. What is also interesting to note is that the *AMs* only publish a decision model not the actual investments a broker performed and thus restrict the access to confidential transactions.

7.2   Application of Pocket Data Mining in the Health Sector

The Ageing Agenda is concerned with the increased life expectancy in the United Kingdom [69]. This represents an increasing economic burden. *PDM* could be used for real-time resource management in order to monitor the person's health, and locate and deploy services and medical resources such as nurses most efficiently. The fact that state-of-the-art medical devices like *ECG* monitors and pulse oximeter are now lightweight and *Bluetooth* enabled[1] makes the *PDM* system timely and important.

Sensors of smartphones can collect in situ continuously streamed data about the health condition of a patient. For example, some earphones of mobile phones can read a person's blood pressure [77], the accelerometer could detect physical activity, also the body temperature could be recorded, and the fact that the mobile phone is

---

[1] Alive Technologies: http://www.alivetec.com/

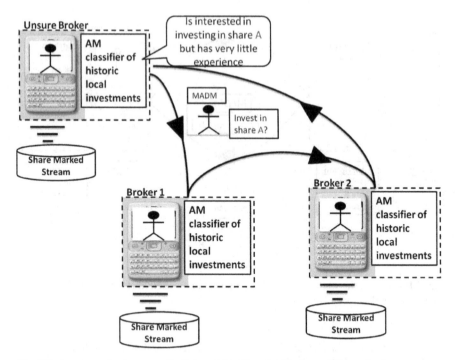

Fig. 7.2 An unsure broker consults the models of local AMs from different brokers

used indicates that the user is conscious and probably well. Behavior patterns can be mined from this sensory data. Nurses and 'mobile medical staff' could be equipped with mobile devices as well, if the nurse is idle, then the mobile phone can send out a mobile agent that roams the network of patients and makes a decision where the nurse is needed the most and instructs the nurse to go there. This decision may take the health status, the location of the patient and the nurse into account as well as the nurse's particular area of expertise.

A typical scenario for *PDM* would include a number of smartphones and/or tablet PCs, and a number of data analysis programs that run on-board these devices. These programs are tailored towards serving the application. The primary motivation for developing this technology is to enable seamless collaboration among users of smartphones. Two constraints necessitate the distribution of the task resulting in a collaborative environment. First, potentially large amounts of medical sensor data that challenge the state-of-the-art of our smartphones. Second, different smartphones may be used by specialists that potentially collect, sense and record different data in the same application area. Collaborative mining addresses these constraints realizing the potential of *PDM*.

The fact that several *MADM*s could be run in parallel on different mobile devices will speed up the data mining/analysis process. Fusing the results of the analysis of different symptoms of a particular patient in real-time would significantly help

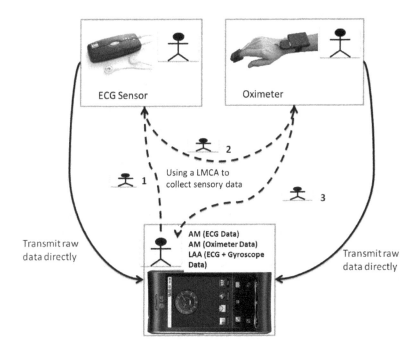

Fig. 7.3 Pocket Data Mining system in the Health Sector: Local View

elderly to live independently, as their smartphones can act as a mediator with the health service departments, if needed.

There are two aspects of the *PDM* architecture in this health application, the localized data analysis depicted in Figure 7.3 and the summarized results managed by the health services depicted in Figure 7.4. The global analysis is facilitated by the *Mobile Agent Decision Makers (MADMs)* whereas the local analysis is facilitated by *Agent Miners (AMs)*. There are also *Local Mobile Collector Agents (LMCAs)* and *Local Aggregation Agents (LAAs)*.

The *LMCA* and *LAA* agents are similar to the already introduced *MADM* and *AM* agents, however, their differences are highlighted below:

- *LAA*: This agent is similar to a local *AM* agent, with the difference that the *LAA* may build their local analysis on the data stream from several sensors, whereas an *AM* may only use the data from a single sensor. Thus, an *AM* in this application may be specialized to raise an alarm of a single sensor that reports medically critical data, whereas the *LAA* takes the data of several sensors into consideration to raise an alarm. For example, an *AM* may analyze data from earphones that record blood pressure. If the blood pressure is critically high, it may raise an alarm, but it may not raise an alarm if the blood pressure is high, but not critically high. In contrary, the *LAA* may take the blood pressure data and also the gyroscope data into consideration. The *LAA* may also raise an alarm if the blood pressure is high (but not critically) in the case that the gyroscope does not suggest any physical activity. This is because high blood pressure and lack of physical activity may suggest a more serious medical condition such as an infection.

- *LMCA*: This simple agent can be used to periodically or on demand collect recent data from the external sensors. This is an alternative to simply transmitting the data from the external sensors to the smartphone. The reason for using an *LMCA* instead of continuously transmitting data is the battery consumption. Transmitting data from external sensors via *Bluetooth* as an example will drain the battery quickly. If assessments are needed in periodical time steps, it may be more economical in terms of battery consumption to collect from time to time a recent snapshot of the data rather than transmitting all the data continuously to the smartphone.

Please note that it would also be a battery efficient option if the sensors can host *AM*s to evaluate their local data rather than transmitting them to the smartphone, as continuous transmission of raw data would drain the batteries of both the sensors and the smartphone rapidly. However, if the computational capabilities of the sensor do not allow hosting and running *AM*s, then the sensors will transmit the raw data to the smartphone. However, we do not extend the discussion to the possibility of *AM*s hosted on external sensors here further. The *AM*s on the smartphones may decide to contact health services when needed using the smartphone's voice and/or messaging services.

Figure 7.4 represents the global view of the *PDM* architecture in the health application.

The health services manage the whole process. The officer in charge may decide to send some agents out to check on a particular patient, or to collect a summarized analysis report for a group of patients. For this task the *MADM* can be used. If an alert is sent to the health services, triggered by an *AM* or *LAA* agent, a medical practitioner in charge may immediately send out a request using an *MADM* asking for the raw data along with a short history of the past few minutes of the data to be re-directed to the central control at the health services department. The medical practitioner then decides whether the alert is positive or not, based on the continuously sent data. If help is not instantly required, the medical practitioner may decide to send out some advice through the messaging services, or call the patient directly.

For example, heart patients with arrhythmia could have their *ECG* monitored along with other physiological symptoms and activities using *PDM* technology. If any problem arises, the smartphone can send the recent recoded *ECG*, other body measurements, and the user's recognised activity of being sitting, walking, or peforming some other activity to the health services department. To decrease false alarms, *PDM* would be able to fuse the analysis of the different measurements to ensure the reality of the risk. For example, the *ECG* analyser may indicate the possibility of a problem in absence of the user's activity. However, having the activity being recognized as the user is getting up the stairs can avoid having a false alarm.

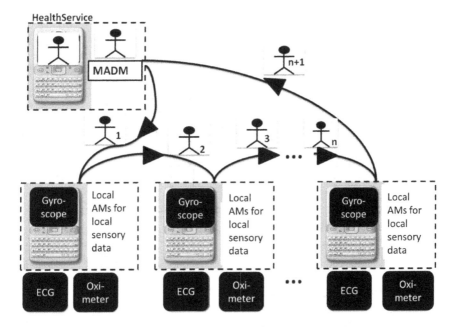

Fig. 7.4 Pocket Data Mining system in the Health Sector: Global View

One may argue that the data could be transmitted continuously from the smartphones and their sensors to the central health services directly. However, this may again result in a communication overhead that drains the batteries of the smartphones rapidly, risking the reliability of the whole system.

7.3   PDM as a Data Fusion System for Decision Support in Public Safety

In this section, we highlight how *PDM* could be used as a data fusion system for decision support for public safety. The basic concept is still based on *PDM* as outlined in Chapter 3. However, rather than employing standard data mining techniques, *PDM* employs additionally data fusion agents. Section 7.3.1 highlights the used infrastructure for this special case of *PDM* followed by example applications in the field of public safety in Sections 7.3.2, 7.3.3, 7.3.4 and 7.3.5. This Section first highlights the data fusion infrastructure and then introduces four scenarios in which the framework could be applied.

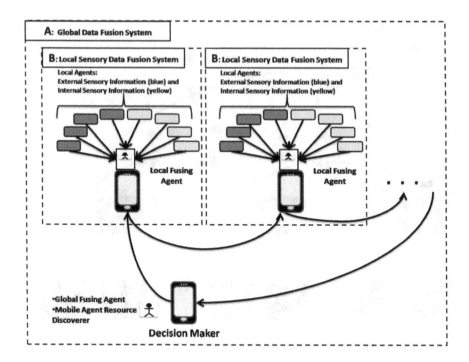

Fig. 7.5 Pocket Data Mining for data fusion and decision support. The system consists of two parts, (A) the Global Data Fusion System and (B) a number of Local Sensory Data Fusion Systems.

7.3.1 PDM as Data Fusion System for Decision Support

Figure 7.5 depicts the data fusion infrastructure consisting of a number of local systems denoted in the figure as B and a global system denoted in the figure as A.

A local Sensory Data Fusion system (B) consists of a single smartphone and several external (blue) and embedded (yellow) data sensors. For example, external sensors could be *ECGs*, oximeters, skin conductors, chemical sensors etc, and embedded sensors could be microphone, gyroscope, thermometer etc. The main component of B is the *Local Fusing Agent (LFA)*. The *LFA* agent is a mobile agent that periodically monitors the collected data from external and the embedded sensors and then fuses the retrieved information intelligently for further analysis and action. The *LFA* agent needs to be situation aware and able to adapt to concept drifts in real-time.

The main components of the Global Data Fusion system (A) are the *Mobile Resource Discoverer agent (MRD)*, which is part of the basic *PDM* system introduced in Chapter 3, and the *Global Data Fusing* agents (*GDF*). The Decision Maker (human) could be located either in the Command Centre or in the field equipped with a smart device. The *MRD* agent can be deployed by the Decision Maker in order to roam the network for the local systems (B). The information collected by the *MRD* agent can be used to generate an itinerary for the *GDF* agent, which is then transferred to the *LFA* agents. The *GDF* agent fuses the information retrieved from the *LFA* agents and returns back to the Decision Maker reporting a meaningful assessment of the current situation in real-time. The *GDF* may adapt its itinerary according to application specific situation changes.

It is important to note that modern smartphones have the capability to act as a hotspot as mentioned in Section 4.4.2 of Chapter 4. Thus, if there is no wireless network, satellite connection or a mobile network available, then each of the smartphones could repeat the same signal, and thus create an ad hoc network. This enables the system to be up and running autonomously even in the absence of a communication network.

This generic infrastructure could be deployed for monitoring and decision support in disaster management, riot management, policing and defence.

7.3.2   Disaster Management and Relief

In case of a disaster, *PDM* could be used to continuously assess the magnitude of the disaster. This could help to predict the future development of the disaster and thus allowing to deploy contingency plan more efficiently.

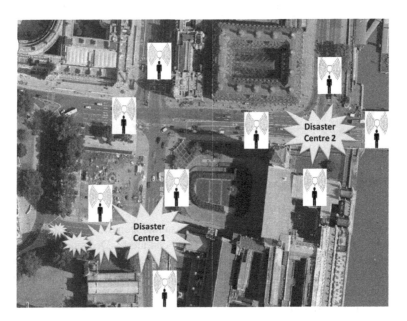

Fig. 7.6 Disaster scenario in an urban area

For example, the scenario outlined in Figure 7.6 represents a critical situation with two disaster centers (e.g., this could be terrorist attacks such as the 7/7 bombings in London, a chemical accident, flooding, or a spreading fire). Assuming the disaster is a number of fires spreading at several locations as shown in Figure 7.6, members of emergency forces close to the disaster can collect relevant data with their smartphones. The captured data then can be fused from a number of smartphones around Disaster Center 1. Processing this information may suggest that the fire is spreading in a certain direction. On the other hand, data fused from smartphones around Disaster Center 2 may indicate that the fire is already contained. Emergency forces can use this live update on the situation for monitoring and decision making on how to deploy the forces most effectively. A similar scenario can easily be drawn for a flooding disaster. Also beneficial in this scenario is that smartphones can build ad hoc networks and do not need a communication network, hence the system will still work even if the communication networks are affected by the disaster.

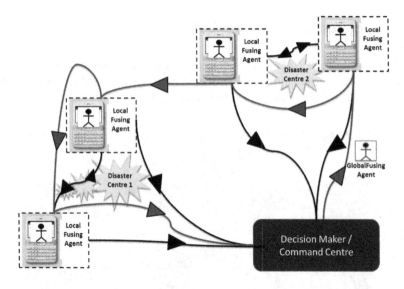

Fig. 7.7 PDM applied to the scenario in Figure 7.6

Figure 7.7 shows how *PDM* could be applied in this scenario. There are two ways of communication: a *Local Fusing Agent (LFA)* can alert a Decision Maker in the field or the Command Centre if a disaster spread is detected on a local device (black arrows). The *LFA* may make use of wireless sensors such as video and temperature sensors, or other for the kind of disaster tailored sensors may be used, such as smoke detectors using *Bluetooth*.

7.3.3 Applications in Riot Management

PDM could be used to assist police and emergency forces to contain social unrest in real-time. There are two possible aspects of the *PDM* system that could be used in such a scenario:

- The use of the smartphones sensing capabilities in order to detect and asses existing riots; and
- the use of social media on smartphones in order to detect concept drifts in the public opinion of potential rioters.

Fig. 7.8 Riot Scenario in an urban area

Figure 7.8 illustrates the usage of *PDM* in the case of riots. There are two basic ways smartphones could be used, first the police forces could use their own smartphones in order to sense detect and evaluate riots and riot potential, and second the smartphones of the rioters and population in the area of interest could be integrated into the system as well.

Similar to the disaster management scenario in Section 7.3.2 police smartphones can be used by police officers in the field in order to detect where rioters are gathering, how many rioters there are, how fast they are moving etc. Again this data could be fused in real-time only using the smartphones and give the police forces an up to date and real-time overview of the situation which will help the police forces to deploy units more effectively.

However, *PDM* could be extended to the smartphones of the population in the affected area. For example the two police units in the scenario may not be able to detect the riots on the right hand side in the scenario. However, information captured by smartphones of civilians in the affected area could be fused into the overall picture of the situation. However, what the smartphone sensors might not detect is the potential for riots. For example, the civilians in the top left corner of Figure 7.8 may be potential rioters. Sentiment analysis of smartphones of civilians, for example sentiment analysis of twitter tweets [6] could reveal that people in this area may be in general unsatisfied and impose a security risk. Based on this information, police forces could be deployed also in areas where social unrest is likely to happen in the future, and thus defuse the situation early.

7.3.4 Applications in Policing

Recent budget cuts of the British coalition government due to the current economic crisis could lead to a reduction of about 60,000 police officers [15]. Reduction in police staff will have to be compensated. *PDM* could help streamlining the process of knowledge acquisition on the crime scene. The crime scene investigators could form an ad hoc network using their smartphones. They could capture pictures, video data and fingerprints as well as any other sensory data on the crime scene or from online data sources (See Figure 7.9). If the task is to know more information about an aspect of the crime, the distribution of tasks could be that one device is to do an Internet search, another is to take pictures and a further one may retrieve data from other sensors close to the crime scene such as *CCTV* cameras. Without *PDM*, the information recorded would have to be formatted and entered into a central data server and a data analyst would have to be employed to evaluate the gathered information, whereas with *PDM* the information could be fused together in real-time to give insights and knowledge about the crime.

7.3.5 Applications in Defence

An important application for smartphones could be real-time assessment of the medical situation of armed forces in the field. This is essentially a hybrid between the health application highlighted in Section 7.2 and the *PDM* data fusion infrastructure discussed in Section 7.3.1. Smartphones of ground personnel could be used to assess the medical condition of individuals as well as of whole units. Medical data could be collected directly by the smartphone of the individual and fused using *LFA* agents. A Command Centre (or a medical officer deployed in the field equipped with smartphone devices), use *MRD* agents and *GDF* agents to fuse the *LFA* agents information. This is in order to assess the medical situation in the field, and to deploy medical resources more efficiently. Any change in the medical condition can be

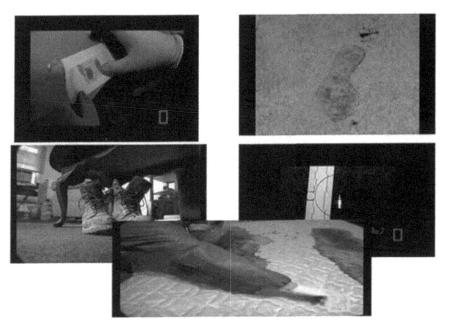

Fig. 7.9 PDM for Crime Scence Investigation

dealt with in real-time supporting effective decision making process. The essential difference to the health application highlighted in Section 7.2 is that the observed personnel are generally healthy soldiers and their health conditions may be more affected by the situation in the field, such as combat injuries. Thus in this scenario it may be recommended to not only just sense the soldier, but also its context such as combat or reconnaissance situations.

Additionally, the soldiers may use their smart devices actively to record information about *Improvised Explosive Devices (IED's)* in order to predict suitable disarming procedures for the discovered *IED*.

7.4   Discussion

This chapter highlights a number of potential applications that can benefit from the *PDM* framework. These applications are by no means exhaustive. Many other applications can benefit from this ambitious project of utilizing the computational power of smartphones and tablet computers. With reports showing that users tend to use these handheld devices putting off their laptops or personal computers[2], *PDM* can serve a number of applications in the era of crowdsensing.

---

[2] http://techcrunch.com/2012/02/06/
when-will-the-post-pc-era-arrive-it-just-did/

Fig. 7.10 Internet of Things – Source *CNN*

In fact, *PDM* can serve a number of important applications when the Internet of things reach its mature stage with high connectivity and seamless access to streaming data generated from the huge number of connected devices [10]. For example, London City airport is the first in the world to adopt the *Internet of Things* technology to solve a number of common problems encountered in airports[3]. Such data can be used by passengers in a *PDM* process to enhance their quality of experience. Figure 7.10 depicts a typical *Internet of Things* network.

---

[3] http://edition.cnn.com//2013/05/02/travel/
london-city-airport-internet-of-things

Chapter 8
Conclusions, Discussion and Future Work

8.1 Summary of Contributions

In this monograph, we detailed the development of our *Pocket Data Mining (PDM)* framework. *PDM* is set to serve the next generation of applications in predictive analytics targeting users of smart handheld devices. We can summarize the presented contribution in this book in the following:

- We presented in Chapter 2, a detailed literature review of developments and research projects related to *PDM*.
- In the core chapter (Chapter 3), *PDM* has been presented giving details of the system architecture and a typical data stream mining process utilizing *PDM*. The chapter also detailed a thorough experimental study evidencing the feasibility of the framework.
- Chapter 4 is a point of reference to practitioners on how *PDM* can be ported from the desktop environment to run on-board *Android* smartphones.
- Chapters 5 and 6 present an important extension to *PDM* to deal with the concept drift issue, which is tightly coupled with change in the user's context.
- Chapter 7 highlights the great potential of *PDM* in a variety of applications in different domains.

With all research projects, there is always room for improvement and extensions. In the following section, current and future work is investigated.

8.2 Ongoing and Future Work

This section provides the reader to pointers to current and future development related to the *PDM* framework.

M.M. Gaber, F. Stahl, and J.B. Gomes, *Pocket Data Mining*, Studies in Big Data 2,
DOI: 10.1007/978-3-319-02711-1_8, © Springer International Publishing Switzerland 2014

8.2.1   Rating System for AMs

The current implementation of the *MADM* agent assumes that the local *AM*s are of good quality, and thus in the case of classification of unlabeled data instances, it is assumed that the weights are calculated correctly and truly reflect the *AM*s classification accuracy. This assumption may be true for the *AM*s we developed in-house, which we used for the evaluation, but third party implementations may not be trusted. For this reason, a rating system about *AM*s is currently being developed based on historical consultations of *AM*s by the *MADM*. For example, if the *MADM* remembers the classifications and weights obtained from *AM*s visited and the true classification of the previously unknown instances is revealed, then the *MADM* could implement its own rating system and rate how reliable an *AM's* weight was in the past. If an *AM* is rated as unreliable, then the *MADM* may even further lower its weight. However, it is essential that this rating system is also able to loosen given ratings, as the *AM's* performance might well change if there is a concept drift in the data stream. In order to detect such concept drifts, it is necessary that *AM*s that have a bad rating are still taken into consideration, even if it is with a low impact due to bad ratings.

We have just outlined a possible rating system for classification *AM*s. However, rating systems for other less generic *PDM* agents such as the *GDF* and *LFA* agents outlined in the applications of *PDM* in Chapter 7 remains an open area to be explored.

8.2.2   Intelligent Schedule for MADMs

In its current implementation, the *MADM* visits all available *AM*s, however, this may be impracticable if the number of *AM*s is very large. Currently, a mechanism is being developed for *MADM*s according to which the *MADM* can decide when to stop consulting further *AM*s. A possible stopping criteria could be that a certain time has elapsed or the classification result is reliable enough. Also the rating system outlined above can be used to determine an order in which *AM*s are visited. If there are time constraints the *MADM* may prioritise more reliable *AM*s.

8.2.3   PDM Beyond Data Stream Classification

PDM is a new niche of distributed data mining. The current implementation of *PDM* focuses on classification techniques, however, there exist many more data mining technologies tailored for data streams and mobile devices. For example, there are stream mining techniques that classify unlabelled data streams [40, 75] which could be introduced into *PDM*.

8.2.4   Resource-Awareness

Resource-aware data mining algorithms as proposed in [39] will boost *PDM*'s applicability in resource constrained mobile networks once integrated in *PDM*. We have briefly reviewed the *Algorithm Granularity* approach to adapting the data stream mining technique to availability of resources and the rate of the incoming streaming data to the smartphone in Chapter 2.

The *MRD* agent can play a role in finding out the initial information needed by the *Algorithm Granularity* approach to set the algorithm parameters. Consequently, a decision can be made on whether the mobile device can deliver results within the set boundaries for quality. Readers can be referred to [34] for more information on the *Algorithm Granularity* approach and its relationship to quality of the produced results.

8.2.5   Visualization of PDM Results

The current version of *PDM* presents the results in a textual form. It is beneficial to develop visualization methods that suit the small screen of handheld devices. Gaber et al [38, 45, 37] proposed a generic approach to data visualization on small devices, termed *Adaptive Clutter Reduction (ACR)*. *ACR* works in a similar way to the *Algorithm Granularity* approach trading off the amount of information presented on the screen to the clarity of this information.

PDM can adopt the *ACR* approach to enhance the graphical user interface of the application. It would also be an interesting extension that *MRD* agent can be used to assess the suitability of the different devices in the *PDM* process according to the size of the screen and the amount of information (data mining output) on each of these devices.

A recent review of visual data analysis methods, including methods for data stream can be found in [94].

8.2.6   New Hardware

Google Glass and smart watches are among devices that reached the market last year[1,2]. Such devices can be utilized by the individual user to serve the different application needs. *PDM* can connect these different devices creating an even smarter environment for the user, analyzing streaming data generated from the different devices.

---

[1] http://www.bbc.co.uk/news/business-23579404

[2] http://www.linkedin.com/influencers/20130806180724-549128-the-coming-revolution-of-wearables

Fig. 8.1 PDM in Action

The opportunities in this area are endless. Personalization of the analysis results is an ultimate goal. In the past, general models were applied, given scarcity
of the data available. With each individual started to generate a high velocity of data
streams with the continuous advances in modern hardware, *PDM* has the potential
for personalization of the data stream mining results.

8.3   Final Words

This monograph style book presents a new technology for data analytics, namely,
the *PDM* framework. With advances in data science and *Big Data* technology, it is
inevitable that the smartphone technology will play an important role in realizing
the full potential of what *Big Data* offers. *PDM* is a pioneering first work that uses
distributed data stream mining in the mobile environment. Adaptations and extensions to this basic framework will continue. The context-aware *PDM* presented in
this monograph is one such extension. However, the framework is extendable and
adaptable to suit the different needs of the targeted applications.

Finally, readers interested in seeing a demonstration of *PDM* are encouraged
to watch the YouTube video whose URL is given in the footnote of this page[3]. A
screenshot form the demonstration is given in Figure 8.1.

---

[3] `http://www.youtube.com/watch?v=MOvlYxmttkE`

References

1. Solution brief: Mobile cloud computing (2010)
2. Hadoop (2011)
3. (2012), http://www.cloudcomputinglive.com/events/170-mobile-cloud-computing-forum.html
4. The 10 largest data bases in the world (2012), http://www.comparebusinessproducts.com/fyi/10-largest-databases-in-the-world
5. Update 2-alibaba launches smartphone running its cloud os (2012), http://in.reuters.com/article/2011/07/28/alibaba-mobile-idINL3E7IS1DF20110728
6. Adedoyin-Olowe, M., Gaber, M.M., Stahl, F.: TRCM: A methodology for temporal analysis of evolving concepts in twitter. In: Rutkowski, L., Korytkowski, M., Scherer, R., Tadeusiewicz, R., Zadeh, L.A., Zurada, J.M. (eds.) ICAISC 2013, Part II. LNCS, vol. 7895, pp. 135–145. Springer, Heidelberg (2013)
7. Aggarwal, C.C., Han, J., Wang, J., Yu, P.S.: A framework for clustering evolving data streams. In: Proceedings of the 29th International Conference on Very Large Data Bases, VLDB 2003, vol. 29. VLDB Endowment (2003), http://dl.acm.org/citation.cfm?id=1315451.1315460
8. Aggarwal, C.C., Han, J., Wang, J., Yu, P.S.: On demand classification of data streams. In: Proceedings of the Tenth ACM SIGKDD International Conference on Knowledge Discovery and Data Mining, KDD 2004, pp. 503–508. ACM, New York (2004), doi:http://doi.acm.org/10.1145/1014052.1014110
9. Agnik: Minefleet description (2011), http://www.agnik.com/minefleet.html
10. Atzori, L., Iera, A., Morabito, G.: The internet of things: A survey. Computer Networks 54(15), 2787–2805 (2010)
11. Bacardit, J., Krasnogor, N.: The infobiotics PSP benchmarks repository. Tech. rep. (2008), http://www.infobiotic.net/PSPbenchmarks
12. Bártolo Gomes, J., Gaber, M.M., Sousa, P.A.C., Menasalvas, E.: Context-aware collaborative data stream mining in ubiquitous devices. In: Gama, J., Bradley, E., Hollmén, J. (eds.) IDA 2011. LNCS, vol. 7014, pp. 22–33. Springer, Heidelberg (2011)
13. Bartolo Gomes, J., Menasalvas, E., Sousa, P.: Learning Recurring Concepts from Data Streams with a Context-aware Ensemble. In: ACM Symposium on Applied Computing, SAC 2011 (2011)

14. Basilico, J.D., Munson, M.A., Kolda, T.G., Dixon, K.R., Kegelmeyer, W.P.: Comet: A recipe for learning and using large ensembles on massive data. CoRR abs/1103.2068 (2011)
15. BBC: Budget cuts 'threaten 60,000 police jobs' (2010), http://www.bbc.co.uk/news/uk-10639938
16. Bellifemine, F.L., Poggi, A., Rimassa, G.: Developing multi-agent systems with JADE. In: Castelfranchi, C., Lespérance, Y. (eds.) ATAL 2000. LNCS (LNAI), vol. 1986, pp. 89–103. Springer, Heidelberg (2001)
17. Berrar, D., Stahl, F., Silva, C.S.G., Rodrigues, J.R., Brito, R.M.M.: Towards data warehousing and mining of protein unfolding simulation data. Journal of Clinical Monitoring and Computing 19, 307–317 (2005)
18. Bifet, A., Gavaldà, R.: Adaptive learning from evolving data streams. In: Adams, N.M., Robardet, C., Siebes, A., Boulicaut, J.-F. (eds.) IDA 2009. LNCS, vol. 5772, pp. 249–260. Springer, Heidelberg (2009)
19. Bifet, A., Holmes, G., Kirkby, R., Pfahringer, B.: Moa: Massive online analysis. The Journal of Machine Learning Research 99, 1601–1604 (2010)
20. Bifet, A., Kirkby, R.: Data stream mining: a practical approach. Tech. rep., Center for Open Source Innovation (2009)
21. Blake, C.L., Merz, C.J.: UCI repository of machine learning databases. Tech. rep., University of California, Irvine, Department of Information and Computer Sciences (1998)
22. Breiman, L.: Bagging predictors. Machine Learning 24(2), 123–140 (1996)
23. Breiman, L.: Random forests. Machine Learning 45(1), 5–32 (2001)
24. Cesario, E., Talia, D.: Distributed data mining patterns and services: an architecture and experiments. In: Concurrency and Computation: Practice and Experience, pp. n/a–n/a (2011), doi:http://dx.doi.org/10.1002/cpe.1877
25. Chan, P., Stolfo, S.J.: Experiments on multistrategy learning by meta learning. In: Proc. Second Intl. Conference on Information and Knowledge Management, pp. 314–323 (1993)
26. Chan, P., Stolfo, S.J.: Meta-Learning for multi strategy and parallel learning. In: Proceedings of the Second International Workshop on Multistrategy Learning, pp. 150–165 (1993)
27. Cortez, P., Lopes, C., Sousa, P., Rocha, M., Rio, M.: Symbiotic Data Mining for Personalized Spam Filtering. In: IEEE/WIC/ACM International Joint Conferences on Web Intelligence and Intelligent Agent Technologies, WI-IAT 2009, vol. 1, pp. 149–156. IEEE (2009)
28. Datta, S., Bhaduri, K., Giannella, C., Wolff, R., Kargupta, H.: Distributed data mining in peer-to-peer networks. IEEE Internet Computing 10(4), 18–26 (2006)
29. Daumé III, H., Marcu, D.: Domain adaptation for statistical classifiers. J. Artif. Int. Res. 26(1), 101–126 (2006), http://dl.acm.org/citation.cfm?id=1622559.1622562
30. Dean, J., Ghemawat, S.: Mapreduce: Simplified data processing on large clusters. Commun. ACM 51, 107–113 (2008)
31. Dinh, H., Lee, C., Niyato, D., Wang, P.: A survey of mobile cloud computing: Architecture, applications, and approaches. In: Wireless Communications and Mobile Computing (2011)
32. Domingos, P., Hulten, G.: Mining high-speed data streams. In: Proceedings of the Sixth ACM SIGKDD International Conference on Knowledge Discovery and Data Mining, KDD 2000, pp. 71–80. ACM, New York (2000), doi:http://doi.acm.org/10.1145/347090.347107

33. Freitas, A.: A survey of parallel data mining. In: Proceedings Second International Conference on the Practical Applications of Knowledge Discovery and Data Mining, London, pp. 287–300 (1998)

34. Gaber, M.: Data stream mining using granularity-based approach. In: Abraham, A., Hassanien, A.E., de Leon, A., de Carvalho, F., Snel, V. (eds.) Foundations of Computational, Intelligence. SCI, vol. 206, pp. 47–66. Springer, Heidelberg (2009), http://dx.doi.org/10.1007/978-3-642-01091-0_3

35. Gaber, M., Krishnaswamy, S., Zaslavsky, A.: Adaptive mining techniques for data streams using algorithm output granularity. In: The Australasian Data Mining Workshop, Citeseer (2003)

36. Gaber, M., Krishnaswamy, S., Zaslavsky, A.: Ubiquitous data stream mining. In: Current Research and Future Directions Workshop Proceedings held in conjunction with The Eighth Pacific-Asia Conference on Knowledge Discovery and Data Mining, Sydney, Australia. Citeseer (2004)

37. Gaber, M.M., Krishnaswamy, S., Gillick, B., AlTaiar, H., Nicoloudis, N., Liono, J., Zaslavsky, A.: Interactive self-adaptive clutter-aware visualisation for mobile data mining. Journal of Computer and System Sciences 79(3), 369–382 (2013), doi:http://dx.doi.org/10.1016/j.jcss.2012.09.009, http://www.sciencedirect.com/science/article/pii/S0022000012001456

38. Gaber, M.M., Krishnaswamy, S., Gillick, B., Nicoloudis, N., Liono, J., AlTaiar, H., Zaslavsky, A.: Adaptive clutter-aware visualization for mobile data stream mining. In: Proceedings of the 2010 22nd IEEE International Conference on Tools with Artificial Intelligence, ICTAI 2010, vol. 2, pp. 304–311. IEEE Computer Society, Washington, DC (2010)

39. Gaber, M.M., Krishnaswamy, S., Zaslavsky, A.: Resource-aware mining of data streams. Journal of Universal Computer Science 11(8), 1440–1453 (2005), http://www.jucs.org/jucs_11_8/resource_aware_mining_of

40. Gaber, M.M., Yu, P.S.: A framework for resource-aware knowledge discovery in data streams: a holistic approach with its application to clustering. In: Proceedings of the 2006 ACM Symposium on Applied Computing, SAC 2006, pp. 649–656. ACM, New York (2006), doi:http://doi.acm.org/10.1145/1141277.1141427

41. Gaber, M.M., Zaslavsky, A., Krishnaswamy, S.: Mining data streams: a review. ACM SIGMOD Record 34, 18–26 (2005), doi:http://doi.acm.org/10.1145/1083784.1083789

42. Gaber, M.M., Zaslavsky, A.B., Krishnaswamy, S.: Data stream mining. In: Data Mining and Knowledge Discovery Handbook, pp. 759–787. Springer (2010)

43. Gantz, J., Reinsel, D.: The digital universe decade, are you ready? In: IDC 2009, pp. 1–16 (2009), http://www.emc.com/digital_universe

44. Garcia, A., Kalva, H.: Cloud transcoding for mobile video content delivery. In: 2011 IEEE International Conference on Consumer Electronics, ICCE, pp. 379–380 (2011), doi:10.1109/ICCE.2011.5722637

45. Gillick, B., AlTaiar, H., Krishnaswamy, S., Liono, J., Nicoloudis, N., Sinha, A., Zaslavsky, A., Gaber, M.M.: Clutter-adaptive visualization for mobile data mining. In: Proceedings of the 2010 IEEE International Conference on Data Mining Workshops, ICDMW 2010, pp. 1381–1384. IEEE Computer Society, Washington, DC (2010)

46. Gomes, J., Krishnaswamy, S., Gaber, M.M., Sousa, P.A., Menasalvas, E.: Mars: a personalised mobile activity recognition system. In: 2012 IEEE 13th International Conference on Mobile Data Management, MDM, pp. 316–319. IEEE (2012)

47. Gomes, J.B., Krishnaswamy, S., Gaber, M.M., Sousa, P.A.C., Menasalvas, E.: Mobile activity recognition using ubiquitous data stream mining. In: Cuzzocrea, A., Dayal, U. (eds.) DaWaK 2012. LNCS, vol. 7448, pp. 130–141. Springer, Heidelberg (2012)
48. Gomes, J.B., Sousa, P.A., Menasalvas, E.: Tracking recurrent concepts using context. Intelligent Data Analysis 16(5), 803–825 (2012)
49. Haghighi, P.D., Krishnaswamy, S., Zaslavsky, A., Gaber, M.M., Sinha, A., Gillick, B.: Open mobile miner: A toolkit for building situation-aware data mining applications. Journal of Organizational Computing and Electronic Commerce (just-accepted, 2013)
50. Hall, M., Frank, E., Holmes, G., Pfahringer, B., Reutemann, P., Witten, I.H.: The weka data mining software: An update. ACM SIGKDD Explorations Newsletter 11(1), 10–18 (2009)
51. Han, J., Kamber, M.: Data Mining: Concepts and Techniques. Morgan Kaufmann (2001)
52. Hillis, W., Steele, L.: Data parallel algorithms. Communications of the ACM 29(12), 1170–1183 (1986)
53. Hmida, M.B.H., Slimani, Y.: Meta-learning in grid-based data mining systems. International Journal of Communication Networks and Distributed Systems 5, 214–228 (2010)
54. Ho, T.K.: Random decision forests. In: International Conference on Document Analysis and Recognition, vol. 1, p. 278 (1995), doi:http://doi.ieeecomputersociety.org/10.1109/ICDAR.1995.598994
55. Hotho, A., Pedersen, R.U., Wurst, M.: Ubiquitous Data. In: May, M., Saitta, L. (eds.) Ubiquitous Knowledge Discovery. LNCS, vol. 6202, pp. 61–74. Springer, Heidelberg (2010)
56. Kargupta, H., Bhargava, R., Liu, K., Powers, M., Blair, P., Bushra, S., Dull, J., Sarkar, K., Klein, M., Vasa, M., Handy, D.: VEDAS: A Mobile and Distributed Data Stream Mining System for Real-time Vehicle Monitoring. In: Proceedings of 2004 SIAM International Conference on Data Mining, SDM 2004, Lake Buena Vista, FL (2004)
57. Kargupta, H., Kargupta, H., Hamzaoglu, I., Hamzaoglu, I., Stafford, B., Stafford, B.: Scalable, distributed data mining using an agent based architecture. In: Proceedings the Third International Conference on the Knowledge Discovery and Data Mining, pp. 211–214. AAAI Press, Menlo Park (1997)
58. Kargupta, H., Hoon Park, B., Pittie, S., Liu, L., Kushraj, D.: Mobimine: Monitoring the stock market from a pda. ACM SIGKDD Explorations 3, 37–46 (2002)
59. Kargupta, H., Puttagunta, V., Klein, M., Sarkar, K.: On-board vehicle data stream monitoring using mine-fleet and fast resource constrained monitoring of correlation matrices. Next Generation Computing, Invited Submission for Special Issue on Learning from Data Streams 25, 5–32 (2007), http://dl.acm.org/citation.cfm?id=1553698.1553700, doi:10.1007/s00354-006-0002-4
60. Katakis, I., Tsoumakas, G., Vlahavas, I.: On the utility of incremental feature selection for the classification of textual data streams. In: Bozanis, P., Houstis, E.N. (eds.) PCI 2005. LNCS, vol. 3746, pp. 338–348. Springer, Heidelberg (2005), http://dx.doi.org/10.1007/11573036_32
61. Kolter, J., Maloof, M.: Dynamic weighted majority: An ensemble method for drifting concepts. The Journal of Machine Learning Research 8, 2755–2790 (2007)
62. Krishnaswamy, S., Gaber, M., Harbach, M., Hugues, C., Sinha, A., Gillick, B., Haghighi, P., Zaslavsky, A.: Open mobile miner: A toolkit for mobile data stream mining. In: Proceedings of the 15th ACM SIGKDD International Conference on Knowledge Discovery and Data Mining (2009), http://eprints.port.ac.uk/4140/

63. Krishnaswamy, S., Loke, S., Zaslasvky, A.: A hybrid model for improving response time in distributed data mining. IEEE Transactions on Systems, Man, and Cybernetics, Part B: Cybernetics 34(6), 2466–2479 (2004), doi:10.1109/TSMCB.2004.836885

64. Langley, P., Iba, W., Thompson, K.: An analysis of bayesian classifiers. In: Proceedings of the Tenth International Conference on Artificial Intelligence, pp. 223–228. MIT Press (1992)

65. Luo, P., Lü, K., Shi, Z., He, Q.: Distributed data mining in grid computing environments. Future Gener. Comput. Syst. 23(1), 84–91 (2007), doi:10.1016/j.future.2006.04.010

66. Masud, M.M., Gao, J., Khan, L., Han, J., Thuraisingham, B.: A practical approach to classify evolving data streams: Training with limited amount of labeled data. In: Proceedings of the 2008 Eighth IEEE International Conference on Data Mining, pp. 929–934. IEEE Computer Society, Washington, DC (2008), http://portal.acm.org/citation.cfm?id=1510528.1511337, doi:10.1109/ICDM.2008.152

67. McClean, B., Hawkins, C., Spagna, A., Lattanzi, M., Lasker, B., Jenkner, H., White, R.: New horizons from multi-wavelength sky surveys. In: Proceedings of the 179th Symposium of the International Astronomical Union held in Baltimore (1998)

68. Miettinen, A.P., Nurminen, J.K.: Energy efficiency of mobile clients in cloud computing. In: Proceedings of the 2nd USENIX Conference on Hot Topics in Cloud Computing, HotCloud 2010, p. 4. USENIX Association, Berkeley (2010)

69. News, B.: Life expectancy rises again, ons says, http://www.bbc.co.uk/news/business-15372869

70. Padovitz, A., Loke, S.W., Zaslavsky, A.: Towards a theory of context spaces. In: Proceedings of the Second IEEE Annual Conference on Pervasive Computing and Communications Workshops, pp. 38–42. IEEE (2004)

71. Page, J., Padovitz, A., Gaber, M.: Mobility in agents, a stumbling or a building block? In: Proceedings of Second International Conference on Intelligent Computing and Information Systems (2005)

72. Page, J., Zaslavsky, A., Indrawan, M., et al.: A buddy model of security for mobile agent communities operating in pervasive scenarios. In: ACM International Conference Proceeding Series, vol. 54, pp. 17–25 (2004)

73. Pan, S., Yang, Q.: A survey on transfer learning. IEEE Transactions on Knowledge and Data Engineering, 1345–1359 (2010)

74. Park, B., Kargupta, H.: Distributed data mining: Algorithms, systems and applications. In: Data Mining Handbook, pp. 341–358. IEA (2002)

75. Phung, N.D., Gaber, M., Rohm, U.: Resource-aware online data mining in wireless sensor networks. In: IEEE Symposium on Computational Intelligence and Data Mining, CIDM 2007, pp. 139–146 (2007), doi:10.1109/CIDM.2007.368865

76. Pittie, S., Kargupta, H., Park, B.H.: Dependency detection in mobimine: A systems perspective. Information Sciences Journal 155(3-4), 227–243 (2003), http://www.sciencedirect.com/science/article/pii/ S0020025503001713, doi:10.1016/S0020-0255(03)00171-3

77. Poh, M.Z., Kim, K., Goessling, A.D., Swenson, N.C., Picard, R.W.: Heartphones: Sensor earphones and mobile application for non-obtrusive health monitoring. In: IEEE International Symposium on Wearable Computers, pp. 153–154 (2009)

78. Provost, F.: Distributed data mining: Scaling up and beyond. In: Advances in Distributed and Parallel Knowledge Discovery, pp. 3–27. MIT Press (2000)

79. Provost, F., Hennessy, D.N.: Distributed machine learning: Scaling up with coarse-grained parallelism. In: Proceedings of the Second International Conference on Intelligent Systems for Molecular Biology, pp. 340–347 (1994)

80. Provost, F., Hennessy, D.N.: Scaling up: Distributed machine learning with cooperation. In: Proceedings of the Thirteenth National Conference on Artificial Intelligence, pp. 74–79. AAAI Press, Menlo Park (1996)

81. Quinlan, R.J.: C4.5: Programs for machine learning. Morgan Kaufmann (1993)

82. Rings, T., Caryer, G., Gallop, J.R., Grabowski, J., Kovacikova, T., Schulz, S., Stokes-Rees, I.: Grid and cloud computing: Opportunities for integration with the next generation network. J. Grid Comput. 7(3), 375–393 (2009)

83. Schlimmer, J., Granger, R.: Beyond incremental processing: Tracking concept drift. In: Proceedings of the Fifth National Conference on Artificial Intelligence, vol. 1, pp. 502–507 (1986)

84. Silva, J.C.D., Giannella, C., Bhargava, R., Kargupta, H., Klusch, M.: Distributed data mining and agents. In Engineering Applications of Artificial Intelligence Journal 18, 791–807 (2005)

85. Stahl, F.: Parallel rule induction. Ph.D. thesis, University of Portsmouth (2009)

86. Stahl, F., Berrar, D., Silva, C.S.G., Rodrigues, J.R., Brito, R.M.M.: Grid warehousing of molecular dynamics protein unfolding data. In: Proceedings of the Fifth IEEE/ACM Int'l Symposium on Cluster Computing and the Grid, pp. 496–503. IEEE/ACM, Cardiff (2005)

87. Stahl, F., Bramer, M.: Random prism: An alternative to random forests. In: Thirty-First SGAI International Conference on Artificial Intelligence, Cambridge, England, pp. 5–18 (2011)

88. Stahl, F., Bramer, M.: Scaling up classification rule induction through parallel processing. Knowledge Engineering Review (in Press)

89. Stahl, F., Bramer, M., Adda, M.: PMCRI: A parallel modular classification rule induction framework. In: Perner, P. (ed.) MLDM 2009. LNCS, vol. 5632, pp. 148–162. Springer, Heidelberg (2009)

90. Stahl, F., Bramer, M., Adda, M.: J-PMCRI: A methodology for inducing pre-pruned modular classification rules. In: Bramer, M. (ed.) IFIP AI 2010. IFIP AICT, vol. 331, pp. 47–56. Springer, Heidelberg (2010)

91. Stahl, F., Gaber, M., Bramer, M.: Scaling up data mining techniques to large datasets using parallel and distributed processing. In: Rausch, P., Sheta, A.F., Ayesh, A. (eds.) Business Intelligence and Performance Management, Advanced Information and Knowledge Processing, pp. 243–259. Springer, London (2013), doi:http://dx.doi.org/10.1007/978-1-4471-4866-1_16

92. Stahl, F., Gaber, M., Bramer, M., Yu, P.: Pocket data mining: Towards collaborative data mining in mobile computing environments. In: 22nd IEEE International Conference on Tools with Artificial Intelligence (ICTAI), vol. 2, pp. 323–330 (2010), doi:10.1109/ICTAI.2010.118

93. Stahl, F., Gaber, M., Bramer, M., Yu, P.: Pocket Data Mining: Towards Collaborative Data Mining in Mobile Computing Environments. In: 2010 22nd IEEE International Conference on Tools with Artificial Intelligence, ICTAI, vol. 2, pp. 323–330. IEEE (2010)

94. Stahl, F., Gabrys, B., Gaber, M.M., Berendsen, M.: An overview of interactive visual data mining techniques for knowledge discovery. Wiley Interdisciplinary Reviews: Data Mining and Knowledge Discovery 3(4), 239–256 (2013), doi:http://dx.doi.org/10.1002/widm.1093

95. Stahl, F., May, D., Bramer, M.: Parallel random prism: A computationally efficient ensemble learner for classification. In: Bramer, M., Petridis, M. (eds.) Research and Development in Intelligent Systems XXIX, pp. 21–34. Springer, London (2012)

96. Street, W., Kim, Y.: A streaming ensemble algorithm (SEA) for large-scale classification. In: Proceedings of the Seventh ACM SIGKDD International Conference on Knowledge Discovery and Data Mining, pp. 377–382. ACM, New York (2001)

97. Szalay, A.: The Evolving Universe. ASSL 231 (1998)

98. Tsymbal, A.: The problem of concept drift: Definitions and related work. In: Computer Science Department, Trinity College Dublin (2004)

99. Tsymbal, A., Pechenizkiy, M., Cunningham, P., Puuronen, S.: Dynamic integration of classifiers for handling concept drift. Inf. Fusion 9, 56–68 (2008), http://portal.acm.org/citation.cfm?id=1297420.1297577, doi:10.1016/j.inffus.2006.11.002

100. Wang, H., Fan, W., Yu, P., Han, J.: Mining concept-drifting data streams using ensemble classifiers. In: Proceedings of the Ninth ACM SIGKDD International Conference on Knowledge Discovery and Data Mining, pp. 226–235. ACM, New York (2003)

101. Way, J., Smith, E.A.: The evolution of synthetic aperture radar systems and their progression to the eos sar. IEEE Transactions on Geoscience and Remote Sensing 29(6), 962–985 (1991)

102. Widmer, G., Kubat, M.: Learning in the presence of concept drift and hidden contexts. Machine Learning 23(1), 69–101 (1996)

103. Witten, I., Frank, E., Hall, M.: Data Mining: Practical Machine Learning Tools and Techniques: Practical Machine Learning Tools and Techniques. The Morgan Kaufmann Series in Data Management Systems. Elsevier Science (2011)

104. Wooldridge, M., Jennings, N.R., et al.: Intelligent agents: Theory and practice. Knowledge Engineering Review 10(2), 115–152 (1995)

105. Wurst, M., Morik, K.: Distributed feature extraction in a p2p setting–a case study. Future Generation Computer Systems 23(1), 69–75 (2007)

106. zASLAVSKY, A.: Mobile agents: Can they assist with context awareness? In: Proceedings of the 2004 IEEE International Conference on Mobile Data Management, pp. 304–305 (2004), doi:10.1109/MDM.2004.1263080

107. Zhu, X., Wu, X., Yang, Y.: Effective classification of noisy data streams with attribute-oriented dynamic classifier selection. Knowl. Inf. Syst. 9, 339–363 (2006), doi:http://dx.doi.org/10.1007/s10115-005-0212-y

Index

Printed in the United States
By Bookmasters